Springer Proceedings in Mathematics & Statistics

Volume 141

Springer Proceedings in Mathematics & Statistics

This book series features volumes composed of selected contributions from workshops and conferences in all areas of current research in mathematics and statistics, including operation research and optimization. In addition to an overall evaluation of the interest, scientific quality, and timeliness of each proposal at the hands of the publisher, individual contributions are all refereed to the high quality standards of leading journals in the field. Thus, this series provides the research community with well-edited, authoritative reports on developments in the most exciting areas of mathematical and statistical research today.

More information about this series at http://www.springer.com/series/10533

Umberto Cherubini · Fabrizio Durante
Sabrina Mulinacci
Editors

Marshall–Olkin Distributions - Advances in Theory and Applications

Bologna, Italy, October 2013

Editors
Umberto Cherubini
Department of Statistics
University of Bologna
Bologna
Italy

Sabrina Mulinacci
Department of Statistics
University of Bologna
Bologna
Italy

Fabrizio Durante
Faculty of Economics and Management
Free University of Bozen-Bolzano
Bolzano
Italy

ISSN 2194-1009 ISSN 2194-1017 (electronic)
Springer Proceedings in Mathematics & Statistics
ISBN 978-3-319-38448-1 ISBN 978-3-319-19039-6 (eBook)
DOI 10.1007/978-3-319-19039-6

Mathematics Subject Classification: 60E05, 60E15, 62H10, 91B30, 91B70, 62P05

Springer Cham Heidelberg New York Dordrecht London

Softcover reprint of the hardcover 1st edition 2015

Printed on acid-free paper

Springer International Publishing AG Switzerland is part of Springer Science+Business Media (www.springer.com)

Preface

Systemic risk is a key problem of this century. One of the most interesting methodological solutions to this problem was given in the past century in the work of Albert Marshall and Ingram Olkin. This book collects advances in this theory presented at the international conference opening the academic year of the Graduate Course in Quantitative Finance at the University of Bologna, held in Bologna in October 2013.

A systemic event is something that affects a set of objects at the same time. With society and the economy becoming global and with unregulated industrial development leading to more extreme kinds of risk, the relevance of systemic risk has dramatically increased. Air pollution and dangerous industrial waste represent a common factor affecting life expectancy of individuals, particularly in specific geographic regions or clusters of population in less developing countries. By the same token, the developments in medical sciences provide a common factor responsible for longer expected lives, particularly in the developed world. Moreover, all of this translates into common risk factors for the insurance sectors, for financial intermediaries, for firms and the society as a whole.

For these reasons the celebrated Marshall–Olkin model, published in 1967, is one of the best tools to address the analysis of risk in this century, the concept of risk being understood in its widest meaning: from that of stopping a machine to that of ending human lives, from natural catastrophes destroying everything built on a region of land to financial catastrophes triggering the default of clusters of firms or banks.

The kernel of the idea is very simple. There is an event that may kill me, one that may kill you, and one that could kill both of us at the same time. Even this raw idea, without any more structure, raises important questions, taking us beyond the technicalities of the model. The only fact that both you and me are exposed to a common shock induces dependence between our lives, because there are scenarios with positive probability in which we die together. This is a special kind of dependence. It is dependence that does not depend on any of us. It is a kind of dependence of which neither you or I carry the blame. It is a sort of background dependence that links our lives because we are located in the same region, we breathe the same air, we work in the same firm, fly on the same airplane. Our lives are dependent because we share the same exposure to some act of the Diabolic Mrs. Nature. For instance, we are exposed to the same catastrophe that may occur in our common region; we are exposed to the same disease carried by the infectious air we are breathing around the same chimney; we are exposed to lay-off if the firm in which we work closes down; we are exposed to die at the same time if the airplane in which we both travel is going to crash. Therefore, systemic risk is all about a background risk changing while we move in space and time, but that in general has moved forward to the front of the scene in the current century.

The structure imposed on the original Marshall–Olkin model was as soft as possible, and this would highlight even more this kind of irresponsible dependence. The shocks killing individuals and that killing all of them were assumed to be independent. They were all assumed to be generated by processes with lack of memory, which are processes for which the past does not have any impact on the lifetime expected in the future. Marshall and Olkin found that in their model this property would carry over to the elements of the cluster. For each of them the intensity, that is, the instantaneous probability of event occurrence to one element in the cluster, would be simply decomposed as the sum of the intensity of the systemic event and that of the idiosyncratic event. While the invariance of the lack of memory property is an important element of the model, which has raised curiosity and discussion among mathematicians, the idea of decomposition of the intensity is also prone to just the opposite need: the possibility to model ageing effects in a flexible way. In fact, invariance of the intensity of life ending is not common in nature, and it would be better to say that it is more the exception than the rule. However, even from the point of view of ageing, the linear decomposition allows an important range of flexibility. For example, in life insurance, the probability of death of an individual could be modelled by limiting ageing to the idiosyncratic part of intensity while keeping invariant the systemic intensity part, or even allowing for a reverse ageing effect common to all individuals and due to the innovation process in medical and healthcare sciences.

It is well known that the best way to provide flexibility to a multivariate model is to extract the dependence function, also known as the copula function, so leaving full flexibility to design the marginal distributions. Even from this point of view, the Marshall–Olkin distribution delivered a very particular, and widely used, copula function. On top of other properties, which we do not discuss here, the main peculiarity is that it provides a dependence function that has both an absolutely continuous and a singular part. In plain terms, if we model survival times or any other set of variables, the Marshall–Olkin copula provides a positive probability that the event occurred at exactly the same time, or that the two variables that are being modelled take the same value. So, in a scatter diagram of a copula function there is a locus on which the points become more concentrated, designing a line with probability mass. This is obviously due to the idea of modelling an event that kills all the elements in the cluster, but it also conveys a property that is quite rare in the realm of copula functions. From the seminal work by Marshall and Olkin, a massive stream of literature has developed, trying to address the flaws of the model, mainly due to its very simple structure. Many of the papers included in this book are devoted to extensions of the model in several directions.

The main extensions are reviewed in the paper by Bernard, Fernàndez, Mai, Schenk and Scherer (BFMSS) in this book. This review, starting with the standard Marshall–Olkin model, describes the main alternative strategies to generate the same distribution by also providing novel representations. In general, the model can be extended in three main directions. One of the first issues that were raised about the model is that all shocks, both specific and systemic, are assumed to be independent. We could call this a pure systemic risk model. In the real world, and particularly in human sciences applications, sometimes disasters, that is failure of the whole system, are triggered by individual failures. A famous solution, reported in the BFMSS review, is the Lévy-frailty model. It is well known that frailty models lead to Archimedean dependence structures. Beyond this benchmark solution, other contributions in this book provide alternative derivations and extensions. Mulinacci proposes a power mixture approach that achieves a similar result of inducing dependence among the shocks in the model. Frostig and Pellerey suggest a dynamics for the common frailty variable. Augusto and Kolev propose a model in which the specific shocks are linked by a dependence structure, while the systemic shock is assumed to be independent.

The model can be extended in a second important direction, with the aim of embedding it in the standard common factor models that we find in linear statistics. The question is that in principle the Marshall–Olkin model, in which the observed variable is the minimum of a idiosyncratic and a systemic variable, is not very different from the standard factor model in which the observed variable is the sum, or an affine function of the two components. Of course, the difference stems from the different variables to which the models are applied. In the Marshall–Olkin model the variables are usually interpreted as lifetimes, are naturally defined on the positive line support, and are assigned an exponential distribution. Instead, in the standard linear factor model application to the stock market the variables are assumed to be returns, defined on the whole line support, and endowed with an elliptical distribution. Therefore, an interesting research question is what results are common to the use of different aggregation functions. This generalization is employed both in the contributed paper by Frostig and Pellerey and that by Durante, Girard and Mazo (DGM). Moreover, the paper by Augusto and Kolev explicitly addresses a natural dual model with respect to the Marshall–Olkin one that arises when the observed variable is the maximum of the idiosyncratic and the common component.

A third line of extension is the dimension of the clusters, namely the number of variables involved. This is the most severe limitation and the most challenging development in the model. In fact, the Marshall–Olkin model suffers from what is called the curse of dimensionality. Increasing the scale of the model very soon induces a degree of combinatorial complexity that is hard to tackle. The solution to the problem is twofold: either we accept to overlook the differences among the individuals and settle for an exchangeable approach or we reduce the number of clusters, by accepting to set to zero some common intensity. In the first case, we preserve the entire spectrum of clusters, but within each one of them, we only consider the average individual. In the second case, we may induce some misspecification in the dependence structure, since the degree of dependence induced by the clusters that are dropped from the analysis may induce a bias in the estimation of dependence in the remaining clusters. Extensions in the direction of non-exchangeability are recalled in the BFMSS paper, and a specific strategy to build models in the Marshall–Olkin spirit, with a focus on estimation issues, is addressed by DGM.

While the contributions in this book are mathematical and statistical in nature, the questions and the extensions addressed have relevant implications for economics, finance and politics. In fact, the economic and financial crisis is the main reason why the concepts of systemic risk and contagion have become paramount in this century. The two concepts have not been actually very well distinguished in public debate. For example, even in the definition of SIFIs, that is systemically important financial institutions, we actually refer to financial entities whose default may bring about a general crisis of the whole system: but this is actually contagion. The effects that followed the default of Lehman Brothers, on 15 September 2008, and that have persisted for years, represent a case of a systemic crisis that was actually triggered by the default of a component of the system.

From the point of view of economic policy and regulation, it is very important to distinguish systemic risk and contagion. Actually, the concept of pure systemic risk coincides with the original Marshall–Olkin framework, in which the common shock affecting all the components of the system is independent of the specific components. It is an act of God of which the components do not take the blame. For this reason, it is quite natural that the effects of this kind of events should be borne out by the community. In other words, a pure systemic event in the Marshall–Olkin spirit is like a natural catastrophe that the community is called to face.

Contagion is different: it is when the collapse of a component of the system brings about the default of a set of other components. Then, the Marshall–Olkin model extensions that drop the independence assumption of the shocks introduce an element of contagion. Each component may play a role in the default of the system as a whole, namely every component may trigger a systemic crisis. From a regulatory point of view, it is quite clear that the natural conclusion is different from that of a pure systemic risk event. More precisely, the expected cost of a systemic crisis triggered by one of the components should be borne out by the components themselves. They must be taxed, instead of the community, to make sure that they provide insurance for the damage that they may cause to the community. It is the principle known as "polluter must pay". Of course, non-exchangeability is also a paramount feature to allow for the application of the model. Consider again the problem that is more fashionable in these days, namely what is the impact of default of a single bank in the rest of the financial system. In addition, what about a major non-financial corporate entity? How do these events change the dependence structure of the remaining risks in the system? For some of these questions we have empirical evidence or at least case studies.

We have seen how the Lehman case was a disaster, bringing about contagion both in the US market and overseas, and not only to the rest of the financial system, but also to the sovereign entities. We also witnessed cases in which a credit event of a large corporate, such as the downgrading of General Motors below the investment grade rating line, was associated to a strong decrease in default dependence. In a pure systemic risk world, all these events should leave the rest of the system, and its dependence unaffected. In a world of exchangeable risk, the default of each element of the system would have the same impact both on the default probability of the other elements in the system and on the dependence structure of the remaining elements. Therefore, removing exchangeability from the system can be considered the next step beyond removal of the independence assumption.

Finally, the generalization of aggregation operators applying to idiosyncratic and systemic unobserved components is also an interesting topic to be discussed in practical applications. On the one hand, it would be interesting to discuss the economic meaning of different aggregation functions: when it has to be linear and which other specific shapes it has to take in other different hidden factor decompositions. On the other hand, there is an interesting question about how the aggregation operators linking hidden factors can be composed with aggregation operators applied to the observed variables. So, for example in standard statistics we have linear systems of observed variables that are in turn a linear combination of unobservable linear factors. In risk management, instead, it also well known that different aggregation operators applied to the risk measure, such as for example the Value-at-Risk, responds to different purposes. Therefore, the sum operator leads to the aggregation of the risk measure: one has to compute the risk measure of the convolution of the risk exposures. Differently, using the max aggregator provides an answer to risk capital allocation, and can be applied to studying the trade-off among different exposures.

In the future, it will be very interesting to address these problems for the Marshall–Olkin model and its extensions. The interest mainly stems from the peculiarity of the dependence structure, and the question of how the singularity in it could affect the results. A first interesting contribution in this direction is provided by Fernandez, Mai and Scherer in their contribution in this book. It addresses the computation of the convolution of random variables linked by Marshall–Olkin dependence, in the simple model of exchangeable risks, finding that in this case one can recover analytical formulas for systems of dimension up to four.

In conclusion, we hope that the results reported in this book could represent a step of development in this interesting field of research started by Marshall and Olkin in the early second half of the past century, and so modern with respect to the main issues raised by the globalisation of risks.

In the end, we express our gratitude to all the people involved in the realization of this project. Moving backward, we would like to thank the referees, both internal and external, for their excellent job in reviewing the papers collected in this book. We thank the authors of the contributed papers for their patience and their timeliness to provide the paper. We thank the Graduate Course in Quantitative Finance for providing the funds for the organization of the international conference where the contributed papers were presented, and the students of the course who provided these funds from their fees. We thank Margherita De Rogatis, the course manager, for materially organizing the conference. We thank Sabrina Mulinacci and Fabrizio Durante, who provided the scientific support for the organization of the conference and the collection of this book. A final warm thanks to all the people who actively took part in the conference, providing comments and suggestions, from the start of the tutorial session to the end of the post-processing networking session. It was a very nice example of how research and learning, presenting and teaching are concepts that are necessary to each other and are indissolubly bound together.

Bologna
March 2015

Umberto Cherubini
Program Director
Graduate Course in Quantitative Finance

Contents

Contributors

German Bernhart Lehrstuhl für Finanzmathematik (M13), Technische Universität München, Garching-Hochbrück, Germany

Fabrizio Durante Faculty of Economics and Management, Free University of Bozen-Bolzano, Bolzano, Italy

Lexuri Fernández Lehrstuhl für Finanzmathematik (M13), Technische Universität München, Garching-Hochbrück, Germany

Esther Frostig Department of Statistics, University of Haifa, Haifa, Israel

Stéphane Girard INRIA, Laboratoire Jean Kuntzmann (LJK), Grenoble, France

Nikolai Kolev Department of Statistics, University of São Paulo, São Paulo, Brazil

Jan-Frederik Mai XAIA Investment, München, Germany

Gildas Mazo INRIA, Laboratoire Jean Kuntzmann (LJK), Grenoble, France

Sabrina Mulinacci Department of Statistics, University of Bologna, Bologna, Italy

Franco Pellerey Dipartimento di Scienze Matematiche, Politecnico di Torino, Turin, Italy

Jayme Pinto Department of Statistics, University of São Paulo, São Paulo, Brazil

Steffen Schenk Lehrstuhl für Finanzmathematik (M13), Technische Universität München, Garching-Hochbrück, Germany

Matthias Scherer Lehrstuhl für Finanzmathematik (M13), Technische Universität München, Garching-Hochbrück, Germany

Chapter 1
A Survey of Dynamic Representations and Generalizations of the Marshall–Olkin Distribution

German Bernhart, Lexuri Fernández, Jan-Frederik Mai, Steffen Schenk and Matthias Scherer

Abstract In the classical stochastic representation of the Marshall–Olkin distribution, the components are interpreted as future failure times which are defined as the minimum of independent, exponential arrival times of exogenous shocks. Many applications only require knowledge about the failure times before a given time horizon, i.e. the model is "truncated" at a fixed maturity. Unfortunately, such a truncation is infeasible with the original exogenous shock model, because it is a priori unknown which arrival times of exogenous shocks are relevant and which ones occur after the given time horizon. In this sense, the original model lacks a time-dynamic nature. Fortunately, the characterization in terms of the lack-of-memory property gives rise to several alternative stochastic representations which are consistent with a dynamic viewpoint in the sense that a stochastic simulation works along a time line and can thus be stopped at an arbitrary horizon. Building upon this dynamic viewpoint, some of the alternative representations lead to interesting generalizations of the Marshall–Olkin distribution. The present article surveys the literature in this regard.

G. Bernhart · L. Fernández · S. Schenk · M. Scherer (✉)
Lehrstuhl für Finanzmathematik (M13), Technische Universität München,
Parkring 11, 85748 Garching-Hochbrück, Germany
e-mail: scherer@tum.de

G. Bernhart
e-mail: german.bernhart@tum.de

L. Fernández
e-mail: lexuri.fernandez@tum.de

S. Schenk
e-mail: steffen.schenk@tum.de

J.-F. Mai
XAIA Investment, Sonnenstr. 19, 80331 München, Germany
e-mail: jan-frederik.mai@xaia.com

U. Cherubini et al. (eds.), *Marshall–Olkin Distributions - Advances in Theory and Applications*, Springer Proceedings in Mathematics & Statistics 141,
DOI 10.1007/978-3-319-19039-6_1

1.1 The Classical Construction of the Marshall–Olkin Law

The motivation behind the seminal paper [38] was to lift the univariate exponential law to higher dimensions. Nevertheless, such an extension is by no means unique. In fact, each of the different characterizations of the univariate exponential law might serve as a plausible starting point for multivariate generalizations. The defining property used in [38] is the lack-of-memory property. The authors succeeded to show that for a d-dimensional survival function $\bar{F} = \bar{F}_{1,\ldots,d}$ with k-dimensional marginals $\bar{F}_{i_1,\ldots,i_k}$, the functional equation

$$\bar{F}_{i_1,\ldots,i_k}(x_{i_1}+y,\ldots,x_{i_k}+y) = \bar{F}_{i_1,\ldots,i_k}(x_{i_1},\ldots,x_{i_k})\,\bar{F}_{i_1,\ldots,i_k}(y,\ldots,y), \tag{1.1}$$

postulated for all $1 \leq i_1 < \cdots < i_k \leq d$, $x_{i_1} \geq 0,\ldots,x_{i_k} \geq 0$, $y > 0$, has a unique solution, and the resulting law is the so-called Marshall–Olkin law. Intuitively, (1.1) means that the conditional distribution—given that the subvector $(X_{i_1},\ldots,X_{i_k})$ of components has survived y years—of the remaining lifetimes is the same as the unconditional distribution of this subvector at time zero. Technically speaking, the multivariate lack-of-memory property (1.1) implies the existence of parameters $\lambda_I \geq 0$, $\emptyset \neq I \subseteq \{1,\ldots,d\}$, with $\sum_{I:k\in I}\lambda_I > 0$, $k = 1,\ldots,d$, such that for all $x_1,\ldots,x_d \geq 0$, one has

$$\bar{F}(x_1,\ldots,x_d) = \exp\Big(-\sum_{\emptyset\neq I\subseteq\{1,\ldots,d\}} \lambda_I \max_{i\in I}\{x_i\}\Big). \tag{1.2}$$

Besides this analytical treatment, [38] also presented a stochastic model for a random vector $(X_1,\ldots,X_d)$ with survival function $\bar{F}$, namely

$$X_k := \min\big\{E_I \,\big|\, \emptyset \neq I \subseteq \{1,\ldots,d\},\ k \in I\big\}, \quad k = 1,\ldots,d, \tag{1.3}$$

where for each subset $\emptyset \neq I \subseteq \{1,\ldots,d\}$, E_I is an exponentially distributed random variable with mean $1/\lambda_I$, denoted by $E_I \sim \mathcal{E}(\lambda_I)$. These $2^d - 1$ random variables are independent. Note that some λ_I can be 0, in which case one has $E_I \equiv \infty$ with probability 1. Since one needs to guarantee that $\sum_{I:k\in I}\lambda_I > 0$ for all $k = 1,\ldots,d$, this means that for each $k = 1,\ldots,d$, there is at least one subset $I \subseteq \{1,\ldots,d\}$ containing the index k such that $\lambda_I > 0$. Figure 1.1 illustrates the stochastic model (1.3). In Fig. 1.2, down-right, the three-dimensional scatterplot of a Marshall–Olkin (survival) copula is shown. The two-dimensional plots reflect the projections of it in a plane. Note that the upper scatterplots display non-exchangeable situations while the one, down-left, shows an exchangeable one.

Remark 1.1 (*How to define subfamilies?*) There is one general problem when working with the Marshall–Olkin distribution in dimensions larger than, say, $d = 3$: one is exposed to the curse of dimensionality, i.e. one faces the challenge to reduce the complexity of the general model without losing its interpretation. The general Marshall–Olkin law has $2^d - 1$ parameters and the original fatal-shock model (1.3)

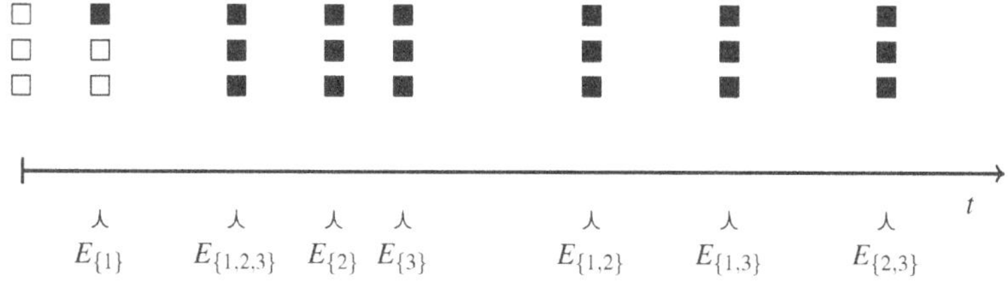

Fig. 1.1 Illustration of the classical Marshall–Olkin model. Three initially functional components (*symbol* □) are hit and destroyed by fatal shocks with arrival times E_I (*symbol* ■). In the above example, we have $X_1 = E_{\{1\}} < X_2 = X_3 = E_{\{1,2,3\}}$

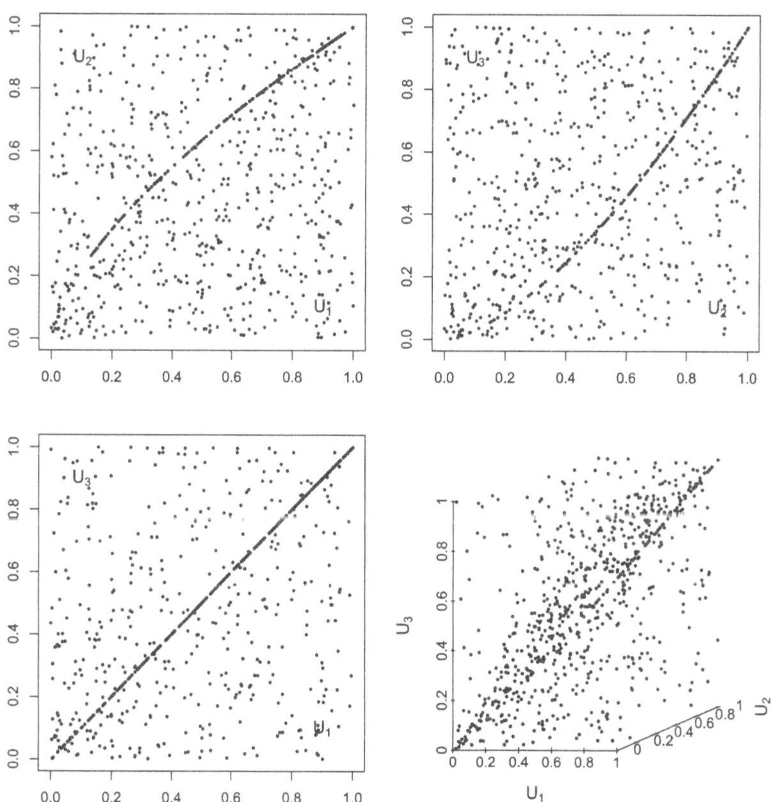

Fig. 1.2 Three-dimensional scatterplot of 750 samples of the Marshall–Olkin (survival) copula with parameters $\lambda_{\{1\}} = 1/10$, $\lambda_{\{2\}} = 1/15$, $\lambda_{\{3\}} = 1/10$, $\lambda_{\{1,2\}} = 1/12$, $\lambda_{\{1,3\}} = 1/12$, $\lambda_{\{2,3\}} = 1/5$, $\lambda_{\{1,2,3\}} = 1/20$ (*down-right*) and the projection of its marginals in two-dimensional scatterplots

has precisely the same number of involved random variables E_I. The consequence is that even seemingly simple tasks, like the computation of probabilities from survival function (1.2) or the simulation of the model via (1.3), become numerically challenging—the effort increasing exponentially in d.

In many real-world situations, one has the advantage that the problem to be modelled itself suggests a natural simplification. For example, some of the shocks cannot occur and thus, the respective rates λ_I can be set to zero, i.e. the random variables $E_I \equiv \infty$ can be omitted. An example in insurance might be certain natural catastrophes that can only cause local damage—say a volcano eruption or a flood. In such a case, it is a modelling task to decide which shocks can be neglected and how the intensities of the remaining ones have to be chosen. Another situation where such a simplification is common practice is portfolio-credit risk, see, [17, 20]. Here, one often considers groupings of loans according to industries and allows only industry sector-specific and idiosyncratic shocks. An even simpler model was used in [11], where only the idiosyncratic shocks $E_{\{k\}}$ and the global shock $E_{\{1,\ldots,d\}}$ are allowed.

Other possibilities to define low-parametric, tractable subfamilies of Marshall–Olkin distributions rely on more convenient stochastic representations in comparison to the original shock model representation (1.3) and are sketched in later sections.

1.1.1 The Exogenous Shock Model Is "Static"

The intuitive nature of the exogenous shock model representation (1.3) has been picked up in numerous applications, e.g. [1, 4, 24, 43], and the E_I are interpreted as future times at which exogenous shocks affect the components of a d-dimensional system. Thus the components themselves are interpreted as times, namely as the failure times of the system's components. In such applications, it is natural to think of the given system in a timely fashion: all components are working at time $t = 0$, then we let the time parameter t run and collect all failure times on the way. Mathematically, one considers the stochastic process $t \mapsto \boldsymbol{X}_t := (\min\{X_1, t\}, \ldots, \min\{X_d, t\})$, $t \geq 0$, which tends to $\boldsymbol{X}_\infty = (X_1, \ldots, X_d)$ as $t \to \infty$. Indeed, in many applications one is actually only interested in the distribution of $\boldsymbol{X}_T$ for a given modelling horizon T rather than in the law of $\boldsymbol{X}_\infty$. Unfortunately, the exogenous shock model representation is out of tune with such a dynamic viewpoint in the sense that in order to simulate $\boldsymbol{X}_t$ for an arbitrary $t > 0$ on the probability space (1.3), there is a priori no way to circumvent the (time-consuming) simulation of $\boldsymbol{X}_\infty$. In other words, the simulation somehow works on a reversed timescale in the sense that $\boldsymbol{X}_\infty$ is simulated first and $\boldsymbol{X}_t$ is derived as a function of $\boldsymbol{X}_\infty$ and t.

The present article surveys alternative stochastic representations of the Marshall–Olkin law which respect a timely intuition in the aforementioned sense. Some of these alternative viewpoints naturally lead to the possibility to construct convenient subfamilies of the Marshall–Olkin law or to generalize the model.

1.2 Interpretation via Poisson Processes

One may interpret the random variable $E_I \sim \mathscr{E}(\lambda_I)$ in (1.3) as the first jump time of a Poisson process $N^I = \{N_t^I\}_{t\geq 0}$ with intensity λ_I, i.e.

$$E_I := \inf\{t \geq 0 : N_t^I = 1\},$$

independently for all non-empty subsets I of $\{1, \ldots, d\}$. Consequently,

$$X_k \stackrel{d}{=} \inf\left\{t \geq 0 : \sum_{\emptyset \neq I \subseteq \{1,\ldots,d\}: k \in I} N_t^I = 1\right\}, \quad k = 1, \ldots, d,$$

which introduces a time parameter t into the model and $\stackrel{d}{=}$ means equal in distribution. Based on this model, the simulation of $\boldsymbol{X}_t$ relies on the simulation of $2^d - 1$ independent Poisson processes N^I until time t. This interpretation was provided in the original reference [38]. It was used in [28] to model the arrival times of insurance claims and in [10, 15, 17, 20] to model credit portfolios. Even though this rewriting into Poisson processes is a little artificial and the simulation of $2^d - 1$ independent Poisson processes might still be an inefficient exercise when $d \gg 2$, it serves as a starting point for generalizations, say by using more general counting processes or by considering stochastically dependent counting processes. An extension to shocks that are not immediately fatal is provided in [39].

1.2.1 Generalization to Cox Processes

Replacing the Poisson processes by more general Cox processes is a popular generalization. Since we only require the first jump times E_I of the respective Cox processes, we may focus on a redefinition of the latter in the sequel. One might rewrite these in terms of unit-exponential random variables $\varepsilon_I \sim \mathscr{E}(1)$ via the so-called canonical construction

$$E_I := \frac{\varepsilon_I}{\lambda_I} = \inf\left\{t \geq 0 : \int_0^t \lambda_I ds \geq \varepsilon_I\right\}. \tag{1.4}$$

On first view, this looks unnecessarily complicated, but it has the advantage of a dynamic interpretation: E_I might be understood as the first moment in time, at which the integrated intensity (or cumulative hazard function) exceeds a unit exponential trigger variable. Equation (1.4) can now be generalized by using either a deterministic intensity $t \mapsto \lambda_I(t)$ or even a stochastic one, i.e. a non-negative stochastic process $\{\lambda_{I,t}\}_{t\geq 0}$. The latter approach was used in the context of portfolio-credit risk by [7, 10, 17]. It has to be noted that by using a generalization to non-constant intensities, the lack-of-memory property of E_I and the subsequently defined X_k is lost, i.e. the model breaks out of the Marshall–Olkin cosmos. Advantages of such a generalization consist of, e.g. the possibility to change the marginal laws to other distributions than the exponential one and to introduce non-stationary innovations.

1.3 The Iterative Construction of Barry Arnold

In [2], an alternative construction of the Marshall–Olkin distribution was given. The main idea is as follows: in the original fatal-shock model, the arrival time ε_1 of the very first component (or the first group of components) being destroyed is—due to min-stability—exponentially distributed, i.e.

$$\begin{aligned}\varepsilon_1 &= \min\big\{X_k \,\big|k \in \{1,\ldots,d\}\big\} \\ &= \min\big\{E_I \,\big|\, \emptyset \neq I \subseteq \{1,\ldots,d\}\big\} \sim \mathscr{E}\Big(\sum_{\emptyset \neq I \subseteq \{1,\ldots,d\}} \lambda_I\Big).\end{aligned}$$

Thus starting at time zero with a system of d working components, we can simulate from an $\mathscr{E}\big(\sum_{\emptyset\neq I\subseteq\{1,\ldots,d\}}\lambda_I\big)$-distribution to reach the point in time ε_1 where the first (or several) component collapses. However, now we have to simulate which subset E_I was responsible for this first shock. Again by min-stability, the probability of shock E_I to be the first one among all shocks is given by $\lambda_I/\sum_{\emptyset\neq I\subseteq\{1,\ldots,d\}}\lambda_I$. Doing so for all shocks defines a discrete probability law on the power set of $\{1,\ldots,d\}$ that we can simulate from to decide which components are destroyed. After the first event—by the lack-of-memory property—we can continue iteratively until all components are destroyed. In mathematical terms, this observation by [2] implies that the survival indicator process $\{\boldsymbol{H}_t\}_{t\geq 0}$, where $\boldsymbol{H}_t := \big(\mathbb{1}_{\{X_1>t\}},\ldots,\mathbb{1}_{\{X_d>t\}}\big)$, satisfies

$$\boldsymbol{H}_t = f\Big(\boldsymbol{H}_s, \bigcup_{k=N_s+1}^{N_t} Y_k\Big), \quad 0 \leq s \leq t, \tag{1.5}$$

for a well-known function[1] $f : [0,\infty)^d \times \mathscr{P}(\{1,\ldots,d\}) \to [0,\infty)^d$, with $N = \{N_t\}_{t\geq 0}$ a Poisson process, whose inter-arrival times are $\varepsilon_1, \varepsilon_2, \ldots$ and $Y_1, Y_2, \ldots \subseteq \{1,\ldots,d\}$ independent and identically distributed (i.i.d.) set-valued random variables, independent of the Poisson process N. Figure 1.3 illustrates the model (1.5).

Unlike the original exogenous shock model representation (1.3), the simulation of the alternative stochastic model (1.5) of [2] can be stopped at any time horizon $T > 0$. A simulation of the random vector $\boldsymbol{X}_T$ depends only on the path of N up to time T, as well as a finite list of independent and identically distributed, set-valued random variables $Y_1,\ldots,Y_{N_T}$. It is clear that the computational effort for this

[1] $\mathscr{P}(\{1,\ldots,d\})$ means the power set of $\{1,\ldots,d\}$.

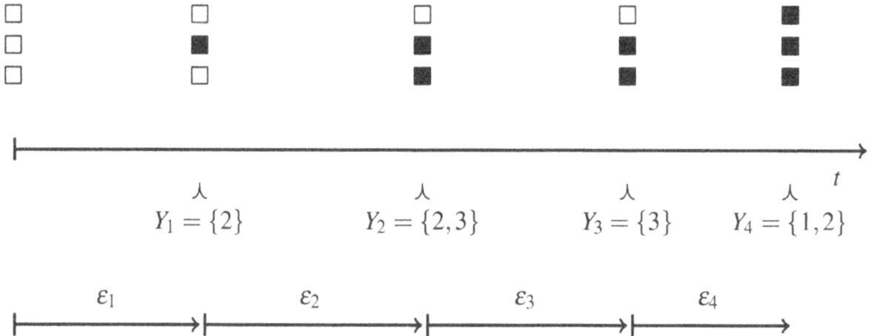

Fig. 1.3 Illustration of Arnold's construction. Three initially functional components (*symbol* □) are hit and destroyed (*symbol* ■) by shocks Y_i that arrive at times $\{\varepsilon_1, \varepsilon_1 + \varepsilon_2, \varepsilon_1 + \varepsilon_2 + \varepsilon_3, \ldots\}$, where the inter-arrival times $\{\varepsilon_i\}_{i \in \mathbb{N}}$ are i.i.d. $\mathscr{E}\big(\sum_{\emptyset \neq I \subseteq \{1,\ldots,d\}} \lambda_I\big)$-distributed. In the above example, we have $X_2 = \varepsilon_1 < X_3 = \varepsilon_1 + \varepsilon_2 < X_1 = \varepsilon_1 + \varepsilon_2 + \varepsilon_3 + \varepsilon_4$

increases in T. Consequently, a simulation of $\boldsymbol{X}_T$ via this model might be preferred over a simulation along the exogenous shock model (1.3) for small time horizons T.

Remark 1.2 (*Simulation of the exchangeable subfamily*) If one is willing to assume that the parameters λ_I of the Marshall–Olkin law depend only on the cardinality $|I|$ of the indexing set I, then one ends up in the exchangeable subfamily, i.e. the law of $(X_{\pi(1)}, \ldots, X_{\pi(d)})$ is invariant with respect to permutations π on $\{1, \ldots, d\}$. It is a d-parametric subfamily, since the subsets I can have cardinality within $\{1, \ldots, d\}$. The reference [32] shows that exchangeable Marshall–Olkin distributions can be simulated with computational effort in $\mathscr{O}(d^2 \log d)$, in contrast to the complexity $\mathscr{O}(2^d)$ of the original exogenous shock model (1.3). Besides exchangeability, the idea of this efficient sampling routine depends heavily on the alternative construction of the Marshall–Olkin distribution of [2].

1.3.1 Generalization to Multivariate Phase-Type Distributions

It is worth mentioning that the stochastic representation (1.5) readily implies that the associated survival indicator process $\{\boldsymbol{H}_t\}_{t \geq 0}$ is a (continuous-time) Markov chain. [9] even characterized the Marshall–Olkin distribution in terms of Markovianity. This alternative viewpoint gives rise to a generalization of Marshall–Olkin distributions to the larger family of multivariate laws associated with random vectors $(X_1, \ldots, X_d)$ whose associated stochastic process $\{\boldsymbol{H}_t\}_{t \geq 0}$ is a Markov chain. These models were studied extensively in [7, 22, 23] in the context of credit-risk modelling. One may even further generalize this family to so-called multivariate phase-type distributions. These distributions arise as absorption times of a continuous-time Markov chain, see

[3, 12, 21] for details. Due to the great level of flexibility of the latter family they are difficult to work with and it seems that they are less popular in applications up to now.

1.4 The Lévy-Frailty Construction

In large dimensions $d \gg 2$, the Marshall–Olkin distribution is challenging to work with, cf. Remark 1.1. In particular, quantities of interest like the average lifetime $A_d := (X_1 + \cdots + X_d)/d$, cf. [19], or the proportion of components which fail before a given time t, i.e. $L_d(t) := \left(\mathbb{1}_{\{X_1 \le t\}} + \cdots + \mathbb{1}_{\{X_d \le t\}}\right)/d$, have a probability distribution which is very difficult to cope with. Generally speaking, these quantities become convenient to handle when the components $X_1, \ldots, X_d$ are conditionally i.i.d., i.e. $X_1, \ldots, X_d$ are i.i.d. conditioned on the σ-algebra generated by the paths of the Lévy subordinator used in construction (1.6) below. In this case, the law of large numbers or the theorem of Glivenko–Cantelli[2] can be applied to compute the limit distribution of A_∞ or $L_\infty(t)$ explicitly, serving as a valid approximation for the large dimension $d \gg 2$ under consideration. This raises the natural question: when is a Marshall–Olkin distribution conditionally i.i.d.?

The answer to this question was found in the dissertation [30], whose main findings were summarized in [29, 31, 33]. It was shown that a Marshall–Olkin distribution is conditionally i.i.d. if and only if there exists a Bernstein function[3] Ψ such that its parameters have a representation via

$$\lambda_I = \sum_{i=0}^{|I|-1} (-1)^i \binom{|I|-1}{i} \left[\Psi(d-|I|+i+1) - \Psi(d-|I|+i)\right], \quad \emptyset \neq I \subseteq \{1, \ldots, d\}.$$

In this case, there exists a Lévy subordinator[4] $S = \{S_t\}_{t \ge 0}$ whose law is uniquely determined by the Laplace transform of its one-dimensional marginals satisfying

$$\mathbb{E}\left[e^{-x\, S_t}\right] = e^{-t\,\Psi(x)}, \quad t, x \ge 0,$$

such that a random vector $(X_1, \ldots, X_d)$ with the Marshall–Olkin distribution in concern can be constructed as

$$X_k := \inf\{t \ge 0 \,:\, S_t \ge \varepsilon_k\}, \quad k = 1, \ldots, d, \tag{1.6}$$

---

[2]See [25], Chap. 5, for these results.

[3]See [41] for details on Bernstein functions.

[4]See [40] for background on Lévy subordinators.

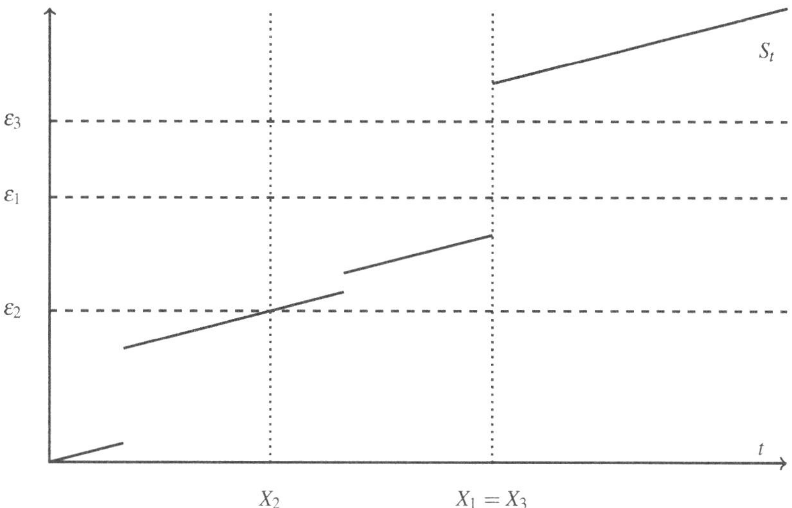

Fig. 1.4 The Lévy-frailty construction of [29]. The components get destroyed once the Lévy subordinator $\{S_t\}_{t\geq 0}$ crosses their respective trigger levels $\varepsilon_1, \ldots, \varepsilon_d$. In this example, component X_2 is hit first, followed by a joint failure of X_1 and X_3. Generalizations of this construction use other increasing stochastic processes

where $\varepsilon_1, \ldots, \varepsilon_d$ is a list of i.i.d. unit exponentials, independent of S. It is apparent from (1.6) that the components $X_1, \ldots, X_d$ are i.i.d. conditioned on (the whole path of) S. More precisely, conditioned on S, the common univariate distribution function of the components is given by $t \mapsto 1 - \exp(-S_t)$, $t \geq 0$. It follows that the aforementioned quantities of interest converge almost surely to

$$A_d \longrightarrow A_\infty \stackrel{d}{=} \int_0^\infty e^{-S_t}\,\mathrm{d}t, \quad L_d(t) \longrightarrow L_\infty(t) \stackrel{d}{=} 1 - e^{-S_t}, \text{ as } d \to \infty.$$

Depending on the chosen Lévy subordinator, these limiting distributions are mathematically tractable. The stochastic model (1.6) is visualized in Fig. 1.4.

1.4.1 Generalizations

The Lévy-frailty construction offers various ways for generalizations of the Marshall–Olkin law. Considering the process $\{L_d(t)\}_{t\geq 0}$ representing the proportion of failed components w.r.t. time, it might be interesting for applications to look at the relative proportion $(L_d(t) - L_d(s))/(1 - L_d(s))$ of failures within the interval $[s, t]$, $0 \leq s < t$. This quantity converges almost surely to

$$\frac{L_d(t) - L_d(s)}{1 - L_d(s)} \longrightarrow \frac{L_\infty(t) - L_\infty(s)}{1 - L_\infty(s)} = 1 - \exp\big(-(S_t - S_s)\big), \quad \text{as } d \to \infty. \tag{1.7}$$

Due to the stationary increments of the Lévy subordinator, this limiting distribution only depends on the length $t - s$ of the interval $[s, t]$ in concern, a feature that can be undesirable in a dynamic environment.

One possibility to relax the Lévy-frailty set-up in this regard is to proceed similarly to the modification discussed in Sect. 1.2.1. The random variables X_k in (1.6) can artificially be rewritten as

$$X_k := \inf\{t \geq 0 : \tilde{S}_t \geq E_k\}, \quad \tilde{S}_t := S_{\int_0^t \lambda_s ds}, \quad k \in \mathbb{N}, \tag{1.8}$$

for a constant intensity $\lambda_t \equiv 1$. A simple and tractable way to introduce randomness now is to insert a random intensity $\{\lambda_t\}_{t\geq 0}$, e.g. modelling $\lambda_t = M$ for all $t \geq 0$, with $M > 0$ a positive random variable. This construction has the nice property that the dependence structure can still be computed analytically, it is of the so-called scale mixture of Marshall–Olkin type as defined, e.g. in [26]. Such a model was used in [6] for the pricing of credit derivatives. However, the resulting dynamics (considering, e.g. (1.7)) are still somewhat artificial.

In contrast, by considering a stochastic process for $\{\lambda_t\}_{t\geq 0}$, it becomes possible to embed the intensity process' parameters and its path progression in the resulting limit in (1.7). An example was given in [34], where the intensity $\{\lambda_t\}_{t\geq 0}$ in (1.8) is the well-known CIR process. For a fixed set of parameters, the Markov property of $\{\lambda_t\}_{t\geq 0}$ implies that the relative loss $1 - \exp\big(-(\tilde{S}_t - \tilde{S}_s)\big)$ depends on the current state λ_s of the intensity process and, thus, changes randomly over time. By incorporating the stochastic time change in the Lévy subordinator, the increments of $\{\tilde{S}_t\}_{t\geq 0}$ neither feature identically distributed nor independent increments. This comes at the cost of losing a vivid interpretation of the dependence structure of $(X_1, \ldots, X_d)$, as the equivalent Marshall–Olkin shock construction in (1.3) does not apply anymore.

A different generalization of the Lévy-frailty construction that maintains the link to an alternative stochastic representation was given in [35]. The authors replace the Lévy subordinator in (1.6) by the more general class of additive processes—the generalization consisting of processes with independent but not necessarily stationary increments. It is shown that the corresponding random vector $(X_1, \ldots, X_d)$ in (1.6) has an alternative representation as in (1.3), the difference being that the independent shocks E_I may have non-exponential distribution functions. Relaxations of such kind can also be found in [16] or [27]. Referring to the dynamic aspect of Eq. (1.7), an additive process $\{S_t\}_{t\geq 0}$ induces that the law of $S_t - S_s$ generally depends on both the starting time point s and the duration $t - s$, yet in a deterministic way. Compared to the generalization involving a stochastic time change, randomness is "lost" while interpretability is analogous to the original Marshall–Olkin model.

Another generalization is given when replacing the Lévy subordinator in (1.6) by the superclass of so-called strong IDT subordinators as defined, e.g. in [37]. It was shown in [36] that this generalization yields min-stable multivariate exponential

(MSMVE) distributions, sometimes also known as min-stable distributions. In fact, these MSMVE laws correspond to an alternative multivariate generalization of the univariate exponential law using min-stability as the defining property, see [18]. These distributions allow for an alternative stochastic representation as well, see, [14], which, however, is not very helpful with respect to simulation as the infimum of a countable sequence of random variables shows up in the corresponding expressions. Furthermore, combining strong IDT subordinators with the previously introduced stochastic time change with the intensity given by a random variable $M > 0$ yields a dependence structure of Archimax type, see [13] for a comprehensive description of this class. Dependence properties of the random vector $(X_1, \ldots, X_d)$ as well as the related default indicators, when a quite arbitrary process is assumed in (1.6) instead of S_t, were investigated in [5]. One possibility to alter the Lévy-frailty set-up, actually not a generalization thereof, is to proceed similarly to the modification discussed in Sect. 1.2.1. For instance, one could consider random variables X_k of the form

$$X_k := \inf\left\{t \geq 0 : \int_0^t \lambda_{k,s} ds \geq \varepsilon_I\right\}, \quad k = 1, \ldots, d,$$

for dependent processes $\{\lambda_{k,s}\}_{s\geq 0}$. In [8] and many of the references therein, $\{\lambda_{k,s}\}_{s\geq 0}$ was modelled as the sum of independent extended CIR processes, where the same summands appear in the construction of various X_k to introduce dependence.

Remark 1.3 (*Non-exchangeable structures/multi-factor models*) We have seen that the (one-factor) Lévy-frailty construction in Eq. (1.6) corresponds to the conditionally i.i.d. subfamily of the Marshall–Olkin law. Similarly, when other increasing processes are used instead of the Lévy subordinator, the resulting law of $(X_1, \ldots, X_d)$ is again conditionally i.i.d., so it is in particular exchangeable. In some applications, however, a richer dependence structure is needed. By starting from a one-factor model, it is actually quite easy to construct more general multi-factor models that give rise to non-exchangeable distributions.

For instance, one idea would be to start with a vector of independent increasing processes $\boldsymbol{S}_t := (S_t^{(1)}, \ldots, S_t^{(n)})'$ and a matrix $A \in \mathbb{R}_+^{d\times n}$. The resulting d-dimensional process $A\,\boldsymbol{S}_t =: \boldsymbol{\Lambda}_t$ is increasing in each coordinate and might be used to generalize construction (1.6) by using the kth coordinate process $\Lambda_t^{(k)}$ as stochastic clock in the definition of X_k instead of S_t. This construction was used, e.g. in [42]. Alternative ideas to define non-exchangeable structures from exchangeable building blocks exploit min-stability properties or a convex combination of subordinators, see [31].

1.5 Conclusion

We surveyed different stochastic representations of the Marshall–Olkin distribution. We put an emphasis on those representations with a dynamic interpretation. Finally, we indicated how different stochastic models served as a basis for different generalizations of the Marshall–Olkin distribution.

References

1. Anastasiadis, S., Chukova, S.: Multivariate insurance models: an overview. Insur.: Math. Econ. **51**, 222–227 (2012)
2. Arnold, B.: A characterization of the exponential distribution by multivariate geometric compounding. Sankhya: Indian J. Stat. **37**, 164–173 (1975)
3. Assas, D., Langberg, N.A., Savits, T.H., Shaked, M.: Multivariate phase-type distributions. Oper. Res. **32**, 688–702 (1984)
4. Baglioni, A., Cherubini, U.: Within and between systemic country risk. Theory and evidence from the sovereign crisis in Europe. J. Econ. Dyn. Control **37**, 1581–1597 (2013)
5. Bäuerle, N., Schmock, U.: Dependence properties of dynamic credit risk models. Stat. Risk Model. Appl. Financ. Insur. **29**, 345–346 (2012)
6. Bernhart, G., Escobar-Anel, M., Mai, J.-F., Scherer, M.: Default models based on scale mixtures of Marshall-Olkin copulas: properties and applications. Metrika **76**, 179–203 (2013)
7. Bielecki, T.R., Crépey, S., Jeanblanc, M., Zargari, B.: Valuation and hedging of CDS counterparty exposure in a Markov copula model. Int. J. Theor. Appl. Financ. **15**, 1–39 (2012)
8. Bielecki, T.R., Cousin, A., Crépey, S., Herbertsson, A.: A bottom-up dynamic model of portfolio credit risk with stochastic intensities and random recoveries. Commun. Stat.—Theory Methods **43**, 1362–1389 (2014)
9. Brigo D., Mai J.-F., Scherer M.: Consistent iterated simulation of multivariate default times: a Markovian indicators characterization. Working paper (2014)
10. Brigo, D., Pallavicini, A., Torresetti, R., Herbertsson, A.: Calibration of CDO tranches with the dynamical generalized-Poisson loss model. Risk **20**, 70–75 (2007)
11. Burtschell, X., Gregory, J., Laurent, J.-P.: A comparative analysis of CDO pricing models. J. Deriv. **16**, 9–37 (2009)
12. Cai, J., Li, H., Laurent, J.-P.: Conditional tail expectations for multivariate phase-type distributions. J. Appl. Probab. **42**, 810–825 (2005)
13. Charpentier A., Fougères A.-L., Genest C., Nešlehová J.-G.: Multivariate Archimax copulas. J. Multivar. Anal. **126**, 118–136 (2014)
14. de Haan, L., Pickands, J.: Stationary min-stable stochastic processes. Probab. Theory Relat. Fields **72**, 477–492 (1986)
15. Duffie, D., Singleton, K.: Simulating Correlated Defaults. Bank of England, London (1998)
16. Durante, F., Quesada-Molina, J.J., Úbeda-Flores, M.: On a family of multivariate copulas for aggregation processes. Inf. Sci. **177**, 5715–5724 (2007)
17. Elouerkhaoui, Y.: Pricing and hedging in a dynamic credit model. Int. J. Theor. Appl. Financ. **10**, 703–731 (2007)
18. Esary, J.D., Marshall, A.W.: Multivariate distributions with exponential minimums. Ann. Stat. **2**, 84–98 (1974)
19. Fernández, L., Mai, J.-F., Scherer, M.: The mean of Marshall-Olkin dependent exponential random variables. In: Cherubini, U., Mulinacci, S., Durante, F. (eds.) Marshall-Olkin Distributions—Advances in Theory and Applications. Springer Proceedings in Mathematics and Statistics. Springer, Switzerland (2015)

20. Giesecke, K.: Correlated default with incomplete information. J. Bank. Financ. **28**, 1521–1545 (2004)
21. Goff, M.: Multivariate discrete phase-type distributions. Ph.D. Thesis. Washington State University (2005)
22. Herbertsson, A., Rootzén, H.: Pricing kth-to-default Swaps under default contagion: the matrix-analytic approach. J. Comput. Financ. **12**, 49–72 (2008)
23. Herbertsson, A.: Modelling default contagion using multivariate phase-type distributions. Rev. Deriv. Res. **14**, 1–36 (2011)
24. Klein, J.-P., Keiding, N., Kamby, C.: Semiparametric Marshall-Olkin models applied to the occurrence of metastases at multiple sites after breast cancer. Biometrics **45**, 1073–1086 (1989)
25. Klenke, A.: Probability Theory: A Comprehensive Course. Springer, Heidelberg (2013)
26. Li, H.: Orthant tail dependence of multivariate extreme value distributions. J. Multivar. Anal. **100**, 243–256 (2009)
27. Li, X., Pellerey, F.: Generalized Marshall-Olkin distributions and related bivariate aging properties. J. Multivar. Anal. **102**, 1399–1409 (2011)
28. Lindskog, F., McNeil, A.J.: Common Poisson shock models: applications to insurance and credit risk modelling. Astin Bull. **33**, 209–238 (2003)
29. Mai, J.-F., Scherer, M.: Lévy-frailty copulas. J. Multivar. Anal. **100**, 1567–1585 (2009)
30. Mai, J.-F.: Extendibility of Marshall-Olkin distributions via Lévy subordinators and an application to portfolio credit risk. Ph.D. Thesis. Technische Universitä München (2010)
31. Mai, J.-F., Scherer, M.: Reparameterizing Marshall-Olkin copulas with applications to sampling. J. Stat. Comput. Simul. **81**, 59–78 (2011)
32. Mai, J.-F., Scherer, M.: Sampling exchangeable and hierarchical Marshall-Olkin distributions. Commun. Stat.-Theory Methods **42**, 619–632 (2013)
33. Mai, J.-F., Scherer, M.: Extendibility of Marshall-Olkin distributions and inverse Pascal triangles. Braz. J. Probab. Stat. **27**, 310–321 (2013)
34. Mai, J.-F., Olivares, P., Schenk, S., Scherer, M.: A multivariate default model with spread and event risk. Appl. Math. Financ. **21**, 51–83 (2014)
35. Mai J.-F., Schenk S., Scherer M.: Exchangeable exogenous shock models. Forthcoming in Bernoulli (2014)
36. Mai, J.-F., Scherer, M.: Characterization of extendible distributions with exponential minima via processes that are infinitely divisible with respect to time. Extremes **17**, 77–95 (2014)
37. Mansuy, R.: On processes which are infinitely divisible with respect to time. In: Cornell University Library. Available via arXiv. http://arxiv.org/abs/http://arxiv.org/abs/math/0504408. Submitted April 2005
38. Marshall, A.W., Olkin, I.: A multivariate exponential distribution. J. Am. Stat. Assoc. **62**, 30–44 (1967)
39. Ross, S.M.: Generalized Poisson shock models. Ann. Probab. **9**, 896–898 (1981)
40. Sato, K.I.: Lévy Processes and Infinitely Divisible Distributions. Cambridge University Press, Cambridge (1999)
41. Schilling, R.L., Song, R., Vondraček, Z.: Bernstein Functions: Theory and Applications. Walter de Gruyter, Berlin (2010)
42. Sun Y., Mendoza-Arriaga R., Linetsky V.: Marshall-Olkin multivariate exponential distributions, multidimensional Lévy subordinators, efficient simulation, and applications to credit risk. Working paper (2013)
43. Wong, D.: Copula from the limit of a multivariate binary model. J. Am. Stat. Assoc. **4**, 35–67 (2000)

Chapter 2
Copulas Based on Marshall–Olkin Machinery

Fabrizio Durante, Stéphane Girard and Gildas Mazo

Abstract We present a general construction principle for copulas that is inspired by the celebrated Marshall–Olkin exponential model. From this general construction method, we derive special subclasses of copulas that could be useful in different situations and recall their main properties. Moreover, we discuss possible estimation strategy for the proposed copulas. The presented results are expected to be useful in the construction of stochastic models for lifetimes (e.g., in reliability theory) or in credit risk models.

2.1 Introduction

The study of multivariate probability distribution function has been one of the classical topics in the statistical literature once it was recognized at large that the independence assumption cannot describe conveniently the behavior of a random system composed by several components. Since then, different attempts have been done in order to provide more flexible methods to describe the variety of dependence types that may occur in practice. Unfortunately, the study of high-dimensional models is not that simple when the dimension goes beyond 2 and the range of these models is still not rich enough for the users to choose one that satisfies all the desired properties.

One of the few examples of high-dimensional models that have been used in an ample spectrum of situations is provided by the Marshall–Olkin distribution,

---

F. Durante (✉)
Faculty of Economics and Management, Free University of Bozen-Bolzano, Bolzano, Italy
e-mail: fabrizio.durante@unibz.it

S. Girard · G. Mazo
Laboratoire Jean Kuntzmann (LJK), INRIA, Grenoble, France
e-mail: stephane.girard@inria.fr

G. Mazo
e-mail: gildas.mazo@inria.fr

U. Cherubini et al. (eds.), *Marshall–Olkin Distributions - Advances in Theory and Applications*, Springer Proceedings in Mathematics & Statistics 141,
DOI 10.1007/978-3-319-19039-6_2

introduced in [24] and, hence, developed through various generalizations (as it can be noticed by reading the other contributions to this volume).

The starting point of the present work is to combine the general idea provided by Marshall–Olkin distributions with a copula-based approach. Specifically, we provide a general construction principle, the so-called Marshall–Olkin machinery, that generates many of the families of copulas that have been recently considered in the literature. The methodology is discussed in detail by means of several illustrations. Moreover, possible fitting strategies for the proposed copulas are also presented.

2.2 Marshall–Olkin Machinery

Consider a system composed by $d \geq 2$ components with a random lifetime. We are mainly interested in deriving an interpretable model for the system supposing that the lifetime of each component may be influenced by adverse factors, commonly indicated as *shocks*. Such shocks can be, for instance, events happening in the environment where the system is working, or simply can be caused by deterioration of one or more components. In a different context, like credit risk, one may think that the system is a portfolio of assets, while the shocks represent arrival times of economic catastrophes influencing the default of one or several assets in the portfolio.

To provide a suitable stochastic model for such situations, let $(\Omega, \mathscr{F}, \mathbb{P})$ be a given probability space.

- For $d \geq 2$, consider the r.v.'s $X_1, \dots, X_d$ such that each X_i is distributed according to a d.f. F_i, $X_i \sim F_i$. Each X_i can be interpreted as a shock that may effect only the ith component of the system, i.e., the idiosyncratic shock.
- Let $\mathscr{S} \neq \emptyset$ be a collection of subsets $S \subseteq \{1, 2, \dots, d\}$ with $|S| \geq 2$, i.e., $\mathscr{S} \subseteq 2^{\{1,2,\dots,d\}}$. For each $S \in \mathscr{S}$ consider the r.v.'s Z_S with probability d.f. G_S. Such a Z_S can be interpreted as an (external) shock that may affect the stochastic behavior of all the system components with index $i \in S$, i.e., the systemic shock.
- Assume a given dependence among the introduced random vectors $\mathbf{X}$ and $\mathbf{Z}$, i.e., suppose the existence of a given copula C such that, according to Sklar's Theorem [32], one has
$$(\mathbf{X}, \mathbf{Z}) \sim C((F_i)_{i=1,\dots,d}, (G_S)_{S \in \mathscr{S}}). \tag{2.1}$$
The copula C describes how the shocks $\mathbf{X}$ and $\mathbf{Z}$ are related each other.
- For $i = 1, \dots, d$, assume the existence of a linking function ψ_i that expresses how the effects produced by the shock X_i and all the shocks Z_S with $i \in S$ are combined together and acts on the ith component.

Given all these assumptions, we call *Marshall–Olkin machinery* any d–dimensional stochastic model $\mathbf{Y} = (Y_1, \dots, Y_d)$ that has the stochastic representation:

$$Y_i = \psi_i(X_i, Z_{S:\, i \in S}). \tag{2.2}$$

Such a general framework includes most of the so-called shock models presented in the literature. Notably, Marshall–Olkin multivariate (exponential) distribution is simply derived from the previous framework by assuming that

$$(\mathbf{X}, \mathbf{Z}) \sim \left(\prod_{i=1}^{d} F_i\right) \cdot \left(\prod_{\emptyset \neq S \in 2^{\{1,2,\dots,d\}}} G_S\right), \tag{2.3}$$

i.e., all the involved r.v.'s are mutually independent, each X_i and each Z_S have exponential survival distribution, $\psi_i = \max$.

However, it includes also various Marshall–Olkin type generalized families, for instance, the family presented in [19] that is obtained by assuming that X_i's are not identically distributed (see also [28]).

By suitable modifications, Marshall–Olkin machinery can be adapted in order to obtain general construction methods for copulas. In fact, the growing use of copulas in applied problems requires the introduction of novel families that may underline special features like tail dependence, asymmetries, etc. Specifically, in order to ensure that the distribution function of $(Y_1, \dots, Y_d)$ of Eq. (2.2) is a copula it could be convenient to select all X_i's and all G_S's with support on $[0, 1]$ and, in addition, ψ_i with range in $[0, 1]$. Obviously, one has also to check that each Y_i is uniformly distributed in $[0, 1]$. We call *Marshall–Olkin machinery* any construction methods for copulas that is based on previous arguments. In the following, we are interested in presenting some specific classes generated by this mechanism.

Provided that the copula C and the marginal d.f.'s of Eq. (2.1) can be easily simulated, distribution functions (in particular, copulas) generated by Marshall–Olkin machinery can be easily simulated. However, if no constraints are require on the choice of $\mathscr{S}$, such distributions are specified by (at least) 2^d parameters, namely

- d parameters related to X_i's;
- $2^d - d - 1$ parameters related to Z_S's;
- (at least) one parameter related to the copula C.

Hence, such a kind of model soon becomes unhandy as the dimension increases. Therefore, we are interested in flexible subclasses generated by Marshall–Olkin machinery with fewer parameters that are better suited for high-dimensional applications.

2.3 Copulas Generated by One Independent Shock

To provide a preliminary class generated by Marshall–Olkin mechanism, consider the case when the system is subjected to individual shocks and one global shock that is independent of the previous ones. In such a case, copulas may be easily obtained in view of the following result.

Theorem 2.1 *For $d \geq 2$, consider the continuous r.v. $\mathbf{X} = (X_1, \ldots, X_d)$ having copula C and such that each X_i is distributed according to a d.f. F supported on $[0, 1]$. Consider the r.v. Z with probability d.f. G such that Z is independent of $\mathbf{X}$. For every $i = 1, \ldots, d$, set*

$$Y_i := \max\{X_i, Z\}.$$

If $G(t) = t/F(t)$ for $t \in]0, 1]$, then the d.f. of $(Y_1, \ldots, Y_d)$ is a copula, given by

$$\widetilde{C}(\mathbf{u}) = G(u_{(1)}) \cdot C(F(u_1), \ldots, F(u_d)), \tag{2.4}$$

where $u_{(1)} = \min_{i=1,\ldots,d} u_i$.

Proof The expression of $\widetilde{C}$ can be obtained by direct calculation. Moreover, since $\widetilde{C}$ is obviously a d.f., the proof consists of showing that the univariate margins of $\widetilde{C}$ are uniform on $(0, 1)$. However, this is a straightforward consequence of the equality $F(t)G(t) = t$ on $(0, 1)$. □

Models of type (2.4) can be also deduced from [29].

Remark 2.1 In the assumption of Theorem 2.1, since G has to be a d.f. it follows that $t \leq F(t)$ for all $t \in [0, 1]$. Moreover, the condition $t/F(t)$ being increasing is equivalent (assuming differentiability of F) to $(\log(t))' \geq (\log(F(t)))'$ on $(0, 1)$. Finally, notice that if F is concave, then $t \mapsto t/F(t)$ is increasing on $(0, 1)$ (see, e.g., [25]).

Remark 2.2 It is worth noticing that the copula $\widetilde{C}$ in (2.4) can be rewritten as

$$\widetilde{C}(\mathbf{u}) = \min(G(u_1), \ldots, G(u_d)) \cdot C(F(u_1), \ldots, F(u_d)).$$

Intuitively, it is the product of the comonotonicity copula $M_d(\mathbf{u}) = \min\{u_1, u_2, \ldots, u_d\}$ and the copula C with some suitable transformation of the respective arguments. This way of combining copulas was considered, for the bivariate case, in [4, 12], and for the general case in [20, Theorem 2.1].

As it can be seen from Fig. 2.1, the main feature of copulas of type (2.4) is that they have a singular component along the main diagonal of the copula domain $[0, 1]^d$. In general, if the r.v. $\mathbf{X}$ has distribution function given by the $\widetilde{C}$ of type (2.4), then $\mathbb{P}(X_1 = X_2 = \cdots = X_d) > 0$. This feature could be of great interest when the major issue is to model a vector of lifetimes and it is desirable that defaults of two or more components may occur at the same time with a nonzero probability.

Roughly speaking, a model of type (2.4) tends to increase the positive dependence. In fact, since F is a d.f. such that $F(t) \geq t$ on $[0, 1]$, $\widetilde{C} \geq C$ pointwise, which corresponds to the positive lower orthant-dependent order between copulas (see, e.g., [16]). However, $\widetilde{C}$ need not be positive dependent, i.e., $C \geq \Pi_d$ pointwise. For instance, consider the random sample from the copula described in Fig. 2.2. As can be noticed, there is no mass probability around the point $(0, 0)$ and, hence, such a copula cannot be greater than Π_2.

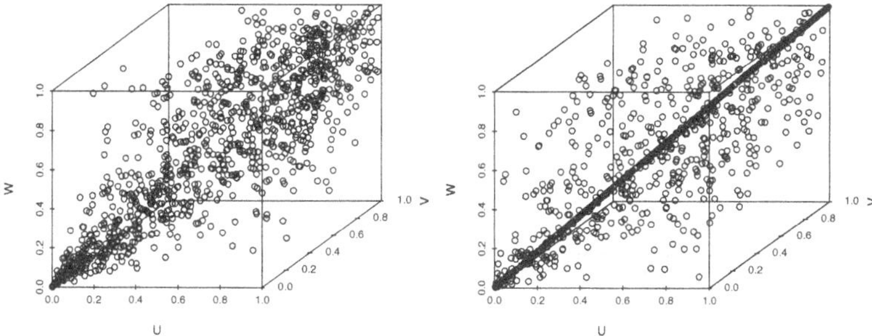

Fig. 2.1 Trivariate Clayton copula (*left*) and its modification of type (2.4) with $F(t) = t^{1-\alpha}$, $\alpha = 0.60$ (*right*)

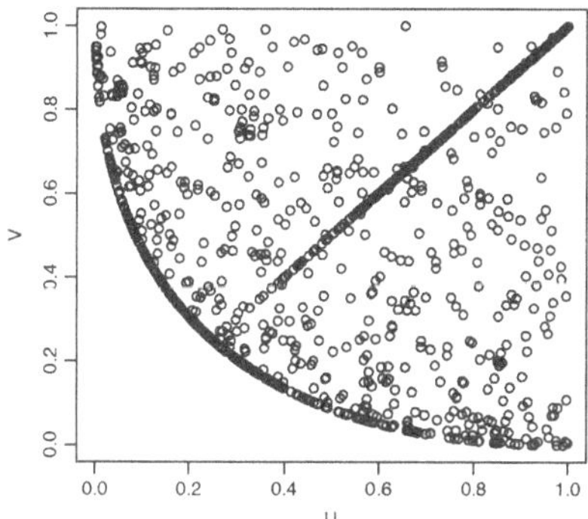

Fig. 2.2 Copula of type (2.4) with one shock generated by $F(t) = t^{1-\alpha}$, $\alpha = 0.50$, and C equal to Fréchet lower bound copula W_2

2.3.1 *The Bivariate Case*

Now, consider the simple bivariate case related to copulas of Theorem 2.1 by assuming, in addition, that C equals the independence copula Π_2. Specifically, we assume that there exist three independent r.v.'s X_1, X_2, Z whose support is contained in $[0, 1]$ such that $X_i \sim F$, $i = 1, 2$, and $Z \sim G(t) = t/F(t)$. For $i = 1, 2$, we define the new stochastic model

$$Y_i = \max(X_i, Z).$$

Then the d.f. of $\mathbf{Y}$ is given by

$$\widetilde{C}(u_1, u_2) = \min(u_1, u_2)F(\max(u_1, u_2)), \tag{2.5}$$

Copulas of this type may be rewritten in the form

$$\widetilde{C}(u_1, u_2) = \min(u_1, u_2)\frac{\delta(\max(u_1, u_2))}{\max(u_1, u_2)} \tag{2.6}$$

where $\delta(t) = \widetilde{C}(t, t)$ is the so-called *diagonal section of* $\widetilde{C}$ (see, for instance, [6]). We refer to [1] for other re-writings. As known, if (U_1, U_2) are distributed according to a copula C, then the diagonal section of C contains the information about the order statistics $\min(U_1, U_2)$ and $\max(U_1, U_2)$. In fact, for every $t \in [0, 1]$

$$\mathbb{P}(\max(U_1, U_2) \leq t) = \delta(t),$$
$$\mathbb{P}(\min(U_1, U_2) \leq t) = 2t - \delta(t).$$

Since the d.f. F related to our shock model equals $\delta(t)/t$ on $(0, 1]$, it follows that it determines the behavior of order statistics of (U_1, U_2). In the case of lifetimes models, this is equivalent to say that the survival of a (bivariate) system is completely driven by one single shock F.

The equivalence of the formulations (2.5) and (2.6) suggests two possible ways for constructing a bivariate model of type (2.4) by either assigning F or δ. In both cases, additional assumptions must be given in order to ensure that the obtained model describes a *bona fide* copula. These conditions are illustrated here (for the proof, see [8]).

Theorem 2.2 *Let* $\widetilde{C}$ *be a function of type (2.4). Set* $\delta := F(t)/t$ *on* $(0, 1]$. *Then* $\widetilde{C}$ *is a copula if, and only if, the functions* $\varphi_\delta, \eta_\delta : (0, 1] \to [0, 1]$ *given by*

$$\varphi_\delta(t) := \frac{\delta(t)}{t}, \qquad \eta_\delta(t) := \frac{\delta(t)}{t^2}$$

are increasing and decreasing, respectively.

Notice that both the independence copula $\Pi_2(u_1, u_2) = u_1 u_2$ and the comonotonicity copula $M_2(u_1, u_2)$ are examples of copulas of type (2.5), generated by $F(t) = t$ and $F(t) = 1$, respectively. Moreover, an algorithm for simulating such copulas is illustrated in [11, Algorithm 1]. Related random samples are depicted in Fig. 2.3. Another example of copulas of type (2.5) is given by the bivariate Sato copula of [21], generated by $F(t) = (2 - t^{1/\alpha})^{-\alpha}$ for every $\alpha > 0$.

Copulas of type (2.5) can be interpreted as the exchangeable (i.e., invariant under permutation of their arguments) members of the family proposed in [23, Proposition 3.1]. Since this latter reference was the first work that has explicitly provided sufficient conditions to obtain copulas of type (2.5), they can also be referred to as *exchangeable Marshall copulas* (shortly, EM copulas), as done in [9]. Notice that EM copulas are also known under the name *semilinear copulas*, a term used in [8], and justified by the fact that these copulas are linear along suitable segments of their domains (see Fig. 2.4).

EM copulas can model positive quadrant dependence, i.e., each EM copula is greater than Π_2 pointwise. Actually, they even satisfy the stronger positive dependence notion called TP2 (see [5]). Following [7] this implies that, if (X, Y) is an exchangeable vector with EM copula, then the vector of residual lifetimes $(X, Y \mid X > t, Y > t)$ at time $t > 0$ is also TP2 and, a fortiori, positive

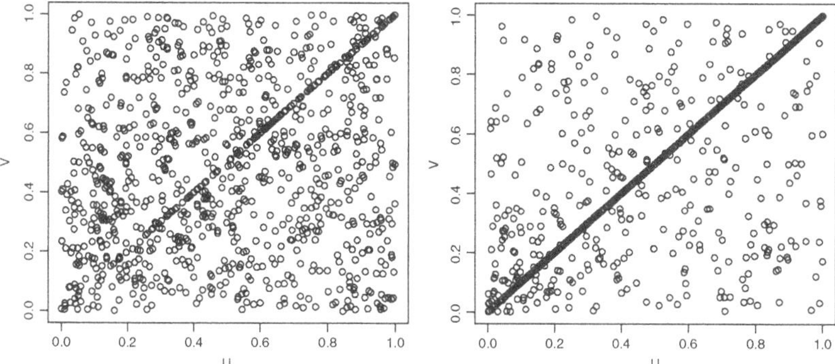

Fig. 2.3 Copulas of type (2.4) generated by $F(t) = t^{1-\alpha}$ with $\alpha = 0.25$ (*left*) and $\alpha = 0.75$ (*right*). These are members of Cuadras–Augé family of copulas

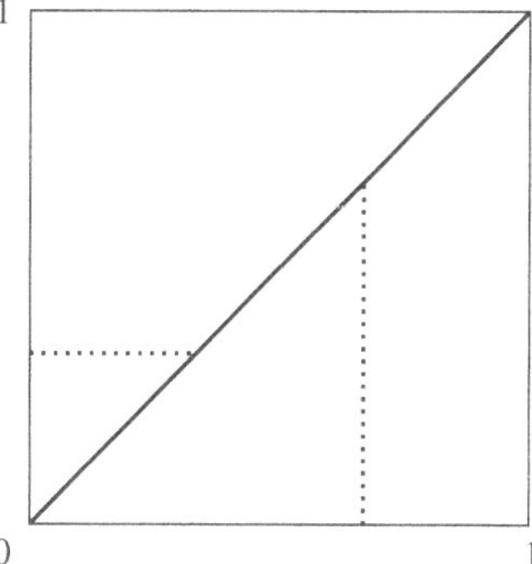

Fig. 2.4 The *dotted lines* indicates the typical segments where the restriction of the EM copula to these sets is linear

quadrant dependent. Roughly speaking, the positive dependence between the residual lifetimes of the system is (qualitatively) preserved at the increase of age.

Another feature of interest in EM copulas is a kind of stability of this class with respect of certain operation. Usually, risk estimation procedures require the calculation of risk functions (like Value-at-Risk) with respect to some specific information about the dependence. In particular, in this respect, upper and lower bounds for copulas with some specified feature are relevant (see, e.g., [2, 17, 33]). Now, the class of all EM copulas is both a convex and log-convex set in the class of all bivariate copulas. Moreover, it is also closed under pointwise suprema and infima operations. Just to provide an example, notice that the class of bivariate Archimedean copulas is neither convex nor closed under suprema and infima.

2.3.2 The Multivariate Case

Copulas of type (2.5) can be easily extended in any dimension. In fact, consider the general case related to copulas of Theorem 2.1 by assuming, in addition, that C equals the independence copula Π_2. Specifically, we assume that there exist $(d+1)$ independent r.v.'s $X_1, X_2, \dots, X_d, Z$ whose support is contained in $[0, 1]$ such that $X_i \sim F$, $i = 1, 2$, and $Z \sim G(t) = t/F(t)$. For $i = 1, 2, \dots, d$, we define the new stochastic model $\mathbf{Y}$, where $Y_i = \max(X_i, Z)$. Then the d.f. of $\mathbf{Y}$ is given by

$$\widetilde{C}(\mathbf{u}) = u_{[1]} \prod_{i=2}^{d} F(u_{[i]}), \tag{2.7}$$

where $u_{[1]}, \dots, u_{[d]}$ denote the components of $(u_1, \dots, u_d)$ rearranged in increasing order. Since $\mathbf{Y}$ has uniform univariate marginals, $\widetilde{C}$ is a copula. Moreover, the following characterization holds (see [10]).

Theorem 2.3 *Let $F\colon [0, 1] \to [0, 1]$ be a continuous d.f., and, for every $d \geq 2$, let $\widetilde{C}$ be the function defined by (2.7). Then $\widetilde{C}$ is a d–copula if, and only if, the function $t \to \frac{F(t)}{t}$ is decreasing on $(0, 1]$.*

Example 2.1 Let $\alpha \in [0, 1]$ and consider $F_\alpha(t) = \alpha t + \overline{\alpha}$, with $\overline{\alpha} := 1 - \alpha$. Then $\widetilde{C}_{F_\alpha}$ of type (2.7) is given by

$$\widetilde{C}_{F_\alpha}(\mathbf{u}) = u_{[1]} \prod_{i=2}^{d} (\alpha u_{[i]} + \overline{\alpha}).$$

In particular, for $d = 2$, we obtain a convex combination of the copulas Π_2 and M_2.

Example 2.2 Let $\alpha \in [0, 1]$ and consider the function $F_\alpha(t) = t^\alpha$. Then $\widetilde{C}_{F_\alpha}$ of type (2.7) is given by

$$\widetilde{C}_{F_\alpha}(\mathbf{u}) = (\min(u_1, u_2, \dots, u_n))^{1-\alpha} \prod_{i=1}^{d} u_i^\alpha.$$

It generalizes the *Cuadras–Augé family* of bivariate copulas [3]. Further generalization of this family is also included in [22].

Copulas of type (2.7) have some distinguished features. First, they are exchangeable, a fact that could represented a limitation in some applications. Second, their tail behavior is only driven by the generator function F. To make this statement precise, consider the following extremal dependence coefficient introduced in [13].

Definition 2.1 Let $\mathbf{X}$ be a random vector with univariate margins $F_1, \dots, F_d$. Let $F_{\min} := \min_i F_i(X_i)$ and $F_{\max} := \max_i F_i(X_i)$. The *lower extremal dependence* *coefficient* (LEDC) and the *upper extremal dependence coefficient* (UEDC) of $\mathbf{X}$ are given, respectively, by

$$\varepsilon_L := \lim_{t\to 0^+} P[F_{\max} \le t | F_{\min} \le t], \quad \varepsilon_U := \lim_{t\to 1^-} P[F_{\min} > t | F_{\max} > t],$$

if the limits exist.

Notice that, in the bivariate case, LEDC and UEDC closely related to the lower and upper tail dependence coefficients (write: LTDC and UTDC, respectively), which are given by

$$\lambda_L = \lim_{t\to 0^+} \frac{C(t,t)}{t} \quad \text{and} \quad \lambda_U = \lim_{t\to 1^-} \frac{1-2t+C(t,t)}{1-t}.$$

The following result holds [10]. Notice that non–trivial LEDC occurs only when F is discontinuous at 0.

Theorem 2.4 *Let $\widetilde{C}$ be a copula of type (2.7) generated by a differentiable F. Then, the LEDC and UEDC are, respectively, given by*

$$\varepsilon_L = \frac{(F(0^+))^{n-1}}{\sum_{i=1}^{n}(-1)^{i-1}\binom{n}{i}(F(0^+))^{i-1}}, \qquad \varepsilon_U = \frac{1-F'(1^-)}{1+(n-1)F'(1^-)}.$$

Although copulas of type (2.7) seem a quite natural generalization of EM copulas, for practical purposes their main inconvenience is that only one function F describes the d–dimensional dependence.

To overcome such oversimplification, a convenient generalization has been provided in [21]. Basically, copulas of type (2.7) have been extended to the form

$$\widetilde{C}(\mathbf{u}) = u_{[1]} \prod_{i=2}^{d} F_i(u_{[i]}), \tag{2.8}$$

for suitable functions $F_2, \ldots, F_d$. Interestingly, a subclass of the considered copulas also can be interpreted in terms of exceedance times of an increasing additive stochastic process across independent exponential trigger variables.

In order to go beyond exchangeable models, keeping a certain tractability and/or simplicity of the involved formulas, a possible strategy could be to combine in a suitable way pairwise copulas of type (2.7) in order to build up a general model. This is pursued, for instance, in [11, 26], by using as building block Marshall–Olkin copulas. Both procedures are described in Sect. 2.4.

2.4 Combining Marshall–Olkin Bivariate Copulas to Get Flexible Multivariate Models

Motivated by the fact that bivariate dependencies are not difficult to check out, it may be of interest to construct a multivariate copula such that each of its bivariate margins depends upon a suitable parameter. For example, following [11], one could introduce a multivariate (extreme-value) copula such that each bivariate marginal C_{ij} belongs to the Cuadras–Augé family:

$$C_{ij}(u_i, u_j) = \Pi_2(u_i, u_j)^{1-\lambda_{ij}} M_2(u_i, u_j)^{\lambda_{ij}}.$$

To this end, following a Marshall–Olkin machinery, one may consider the following stochastic representation of r.v.'s whose support is contained in [0, 1]:

- For $d \geq 2$, consider the r.v.'s $X_1, \dots, X_d$ such that each X_i is distributed according to a d.f. $F_i(t) = t^{1-\sum_{j \neq i} \lambda_{ij}}$ for $i = 1, 2, \dots, d$.
- For $i, j \in \{1, 2, \dots, d\}$, $i < j$, consider the r.v. Z_{ij} distributed according to $G_{ij}(t) = t^{\lambda_{ij}}$.
- We assume independence among all X's and all Z's.

For $i = 1, 2, \dots, d$, we define the new r.v. $\mathbf{Y}$ whose components are given by

$$Y_i = \max\left(X_i, Z_{i1}, \dots, Z_{i(j-1)}, Z_{i(j+1)}, \dots, Z_{id}\right).$$

Basically, Y_i is determined by the interplay among the individual shock X_i and all the pairwise shocks related to the ith component of a system. Then the d.f. of $\mathbf{Y}$ is given by

$$C^{\mathbf{pw}}(\mathbf{u}) = \prod_{i=1}^{d} u_i^{1-\sum_{j \neq i} \lambda_{ij}} \prod_{i<j} (\min\{u_i, u_j\})^{\lambda_{ij}}.$$

Here, for every $i, j \in \{1, \dots, d\}$ and $i < j$, $\lambda_{ij} \in [0, 1]$ and $\lambda_{ij} = \lambda_{ji}$. Moreover, if, for every $i \in \{1, \dots, d\}$, $\sum_{j=1, j \neq i}^{d} \lambda_{ij} \leq 1$, then $C^{\mathbf{pw}}$ is a multivariate d-copula (that is also an extreme–value copula). This copula has Cuadras–Augé bivariate margins and, therefore, may admit nonzero UTDCs. However, even if this model is nonexchangeable, the constraints given on the parameters are a severe drawback. For instance, when $d = 3$, one of the constraint is that $\lambda_{12} + \lambda_{13} \leq 1$. Therefore, the two pairs (X_1, X_2) and (X_1, X_3) cannot have a large UTDC together. In the simplified case where all the parameters λ_{ij}'s are equal to a common value $\lambda \in [0, 1]$, the copula $C^{\mathbf{pw}}$ reduces to

$$C^{\mathbf{pw}}(\mathbf{u}) = \prod_{i=1}^{d} u_{[i]}^{1-\lambda(i-1)}$$

with the constraint $\lambda \leq \frac{1}{d-1}$.

Another way of combining Marshall–Olkin bivariate copulas, that does not suffer from any constraints, and that still yield a flexible model, was hence proposed in [26] and it is given next.

Let $Y_0, Y_1, \ldots, Y_d$ be standard uniform random variables such that the coordinates of $(Y_1, \ldots, Y_d)$ are conditionally independent given Y_0. The variable Y_0 plays the role of a latent, or unobserved, factor. Let us write C_{0i} the distribution of (Y_0, Y_i) and $C_{i|0}(\cdot|u_0)$ the conditional distribution of Y_i given $Y_0 = u_0$, for $i = 1, \ldots, d$. The copulas C_{0i} are called the *linking copulas* because they link the factor Y_0 to the variables of interest Y_i. It is easy to see that the distribution of $(Y_1, \ldots, Y_d)$ is given by the so-called *one-factor copula* [18]

$$C(\mathbf{u}) = \int_0^1 C_{1|0}(u_1|u_0) \ldots C_{d|0}(u_d|u_0)\, du_0. \tag{2.9}$$

When one chooses C_{0i} to be of type (2.5) with generator F_i, the integral (2.9) can be calculated. This permits to exhibit interesting properties for this class of copulas. Thus, calculating the integral yields

$$\begin{aligned} C(\mathbf{u}) = u_{(1)} \Bigg[\left(\prod_{j=2}^{d} u_{(j)} \right) \int_{u_{(d)}}^{1} \prod_{j=1}^{d} F_j'(x)dx + F_{(1)}(u_{(2)}) \left(\prod_{j=2}^{d} F_{(j)}(u_{(j)}) \right) \\ + \sum_{k=3}^{d} \left(\prod_{j=2}^{k-1} u_{(j)} \right) \left(\prod_{j=k}^{d} F_{(j)}(u_{(j)}) \right) \int_{u_{(k-1)}}^{u_{(k)}} \prod_{j=1}^{k-1} F_{(j)}'(x)dx \Bigg], \end{aligned} \tag{2.10}$$

where $F_{(i)} := F_{\sigma(i)}$ and σ is the permutation of $(1, \ldots, d)$ such that $u_{\sigma(i)} = u_{(i)}$. The particularity of this copula lies in the fact that it depends on the generators through their reordering given by the permutation σ. This feature gives its flexibility to the model. Observe also that $C(\mathbf{u})$ writes as $u_{(1)}$ multiplied by a functional of $u_{(2)}, \ldots, u_{(d)}$, form that is similar to (2.7). Interestingly, all the bivariate copulas derived from this model have a simple form as stated below.

Proposition 2.1 *Let C_{ij} be a bivariate margin of (2.10). Then C_{ij} is a copula of type (2.5) with generator*

$$F_{ij}(t) = F_i(t)F_j(t) + t \int_t^1 F_i'(x)F_j'(x)dx.$$

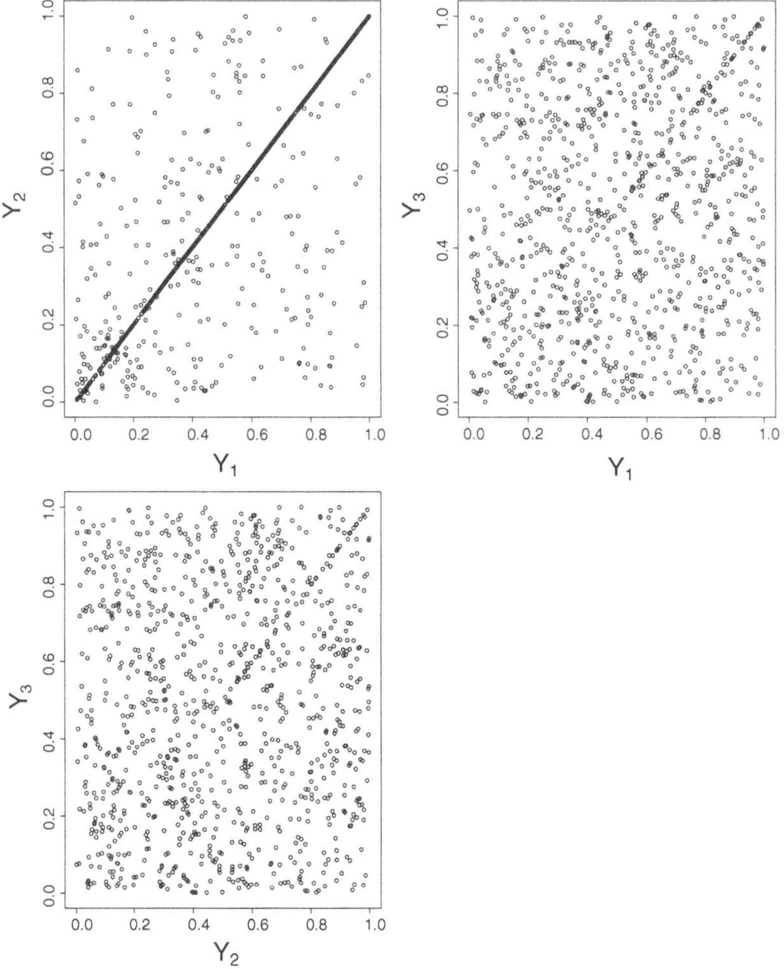

Fig. 2.5 Random sample of 1000 points from a 3–copula of type (2.10) with Cuadras–Augé generators with parameters $(\alpha_1, \alpha_2, \alpha_3) = (0.9, 0.9, 0.1)$. The figure shows the three bivariate margins

By Proposition 2.1, the class of copulas (2.10) can be viewed as a generalization in higher dimension of the bivariate copulas of type (2.5). Moreover, the LTDC and UTDC coefficients are given by

$$\lambda_{L,ij} = F_i(0)F_j(0) \text{ and } \lambda_{U,ij} = (1 - F_i'(1^-))(1 - F_j'(1^-)).$$

Example 2.3 (Fréchet generators) Let $F_i(t) = \alpha_i t + 1 - \alpha_i$, $\alpha_i \in [0, 1]$. By Proposition 2.1, F_{ij} is given by

$$F_{ij}(t) = \big(1 - (1-\alpha_i)(1-\alpha_j)\big)\, t + (1-\alpha_i)(1-\alpha_j).$$

The LTDC and UTDC are, respectively, given by

$$\lambda_{L,ij} = \lambda_{U,ij} = (1-\alpha_i)(1-\alpha_j).$$

Example 2.4 (Cuadras–Augé generators) Let $F_i(t) = t^{\alpha_i}$, $\alpha_i \in [0, 1]$. By Proposition 2.1, F_{ij} is given by

$$F_{ij}(t) = \begin{cases} t^{\alpha_i+\alpha_j}\left(1 - \frac{\alpha_i\alpha_j}{\alpha_i+\alpha_j-1}\right) + t\frac{\alpha_i\alpha_j}{\alpha_i+\alpha_j-1} & \text{if } \alpha_i + \alpha_j \neq 1 \\ t(1-(1-\alpha)\alpha \log t) & \text{if } \alpha = \alpha_j = 1 - \alpha_i. \end{cases}$$

The LTDC and UTDC are, respectively, given by

$$\lambda_{L,ij} = 0 \text{ and } \lambda_{U,ij} = (1-\alpha_i)(1-\alpha_j).$$

In the case $d = 3$, Fig. 2.5 depicts a simulated sample of 1000 observations from this copula with parameter $(\alpha_1, \alpha_2, \alpha_3) = (0.9, 0.9, 0.1)$.

Unlike copulas of type (2.7), the copulas of type (2.10) are not exchangeable. They are determined by d generators $F_1, \dots, F_d$, which combine together to give a more flexible dependence structure. Taking various parametric families, as illustrated in Examples 2.3 and 2.4, allows to obtain various tail dependencies.

2.5 Some Comments About Statistical Inference Procedures

The construction principle presented above provides copulas that are not absolutely continuous (up to trivial cases) with respect to the restriction of the 2-dimensional Lebesgue measure to the copula domain. Thus, statistical procedures that requires density of the related distribution cannot be applied. Moreover, the singular component often implies the presence of points where the derivatives do not exist, a fact that should also be considered for the direct applicability of statistical techniques based on moments' method (see, for instance, [14]).

In this section, we present instead a method to estimate the parameters of the copulas encountered in this paper, which is based on some recent results in [27].

Let

$$(X_1^{(1)}, \dots, X_d^{(1)}), \dots, (X_1^{(n)}, \dots, X_d^{(n)})$$

be a sample of n independent and identically distributed d-variate observations from $(X_1, \dots, X_d)$, a random vector distributed according to F and with copula C, where

$C \equiv C_\theta$ belongs to a parametric family indexed by a parameter vector $\theta \in \Theta \in \mathbb{R}^q$, $q \leq d$. The estimator is defined as

$$\hat{\theta} = \arg\min_{\theta \in \Theta} \left(\hat{r} - r(\theta)\right)^T \widehat{W} \left(\hat{r} - r(\theta)\right), \tag{2.11}$$

where $\hat{r} = (\hat{r}_{1,2}, \ldots, \hat{r}_{d-1,d})$, $r(\theta) = (r_{1,2}(\theta), \ldots, r_{d-1,d}(\theta))$ and $\widehat{W}$ is a positive definite (weight) matrix with full rank; the coordinate $r_{i,j}(\theta)$ is to be replaced by a dependence coefficient between X_i and X_j, and $\hat{r}_{i,j}$ by its empirical estimator—for instance, the Spearman's rho or the Kendall's tau. The approach (2.11) can be viewed as an extension to the multivariate case of the Spearman's rho/Kendall's tau inversion method [15]. The asymptotic properties of $\hat{\theta}$ have been studied in [27] in the case where the copulas do not have partial derivatives on the whole domain as it is the case of the copulas in this article. In particular, it was shown that, under natural identifiability conditions on the copulas, $\hat{\theta}$ exists, is unique with probability tending to 1 as $n \to \infty$, and in that case, is consistent and $\sqrt{n}(\hat{\theta} - \theta)$ tends to a Gaussian distribution.

For the purpose of illustration, we present here a real-data application of the method by using a dataset consisting of 3 gauge stations where annual maximum flood data were recorded in northwestern Apennines and Thyrrhenian Liguria basins (Italy): Airole, Merelli, and Poggi. The dataset is the same used in [11] to which we refer for more detailed description.

In order to fit the dependence among these three gauge stations, we use the class of copulas given by

$$C(u_1, u_2, u_3) = \left(\prod_{i=1}^{3} u_i^{1-\theta_i}\right) \min_{i=1,2,3} (u_i^{\theta_i}), \quad \theta_i \in [0, 1],\ i = 1, 2, 3.$$

Such a copula can be also seen as generated by Marshall–Olkin machinery, by assuming that **X** and **Z** are independent r.v.'s of length d whose copula is given by Π_d and M_d, respectively, F_i and G_i are power functions, and $\psi = \max$.

The estimator (2.11) coordinates $\hat{\theta}_1$, $\hat{\theta}_2$ and $\hat{\theta}_3$ are given by

$$\hat{\theta}_i = \frac{1}{2}\left(1 + \frac{1}{\hat{\tau}_{i,j}} + \frac{1}{\hat{\tau}_{i,k}} - \frac{1}{\hat{\tau}_{j,k}}\right),$$

where the $\hat{\tau}_{i,j}$ are the pairwise sample Kendall's τ coefficients. Notice that, as consequence of [27, Proposition 2], when the number of parameters is equal to the number of pairs $d(d-1)/2$, then the estimator given by (2.11) does not depend on the weights.

These previous estimates help to quantify the critical levels and return periods corresponding to this dataset (see, e.g., [30, 31]). In hydrology, a critical level p corresponding to a return period T is defined through the relationship

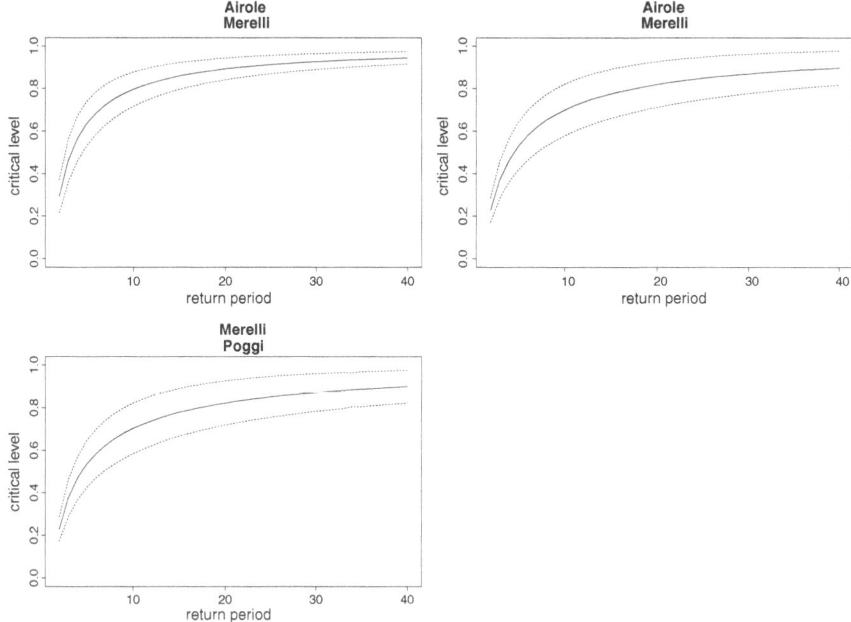

Fig. 2.6 Critical levels for $T = 2, \ldots, 40$ together with 95 % confidence intervals

$$T = \frac{1}{1 - \mathbb{P}(C(F_1(Y_1), \ldots, F_d(Y_d)) \leq p)}, \quad p \in [0, 1],$$

where $Y_1, \ldots, Y_d$ are the r.v.'s of interest and $F_1, \ldots, F_d$ their respective univariate marginals. The return period can be interpreted as the average time elapsing between two dangerous events. For instance, $T = 30$ years means that the event happens once every 30 years in average. Figure 2.6 shows the estimated critical levels, along with confidence intervals, associated to the fitted dataset.

2.6 Conclusions

We have presented a construction principle of copulas that is inspired by the seminal Marshall–Olkin idea of constructing shock models. The copulas obtained in this way have some distinguished properties:

- they have an interpretation in terms of (local or global) shocks;
- they enlarge known families of copulas by including asymmetric copula (in the tails) and/or nonexchangeability;
- they have a natural sampling strategy;

- they can be used to build models with singular components, a fact that is useful when modeling joint defaults of different lifetimes (i.e., credit risk).
- they can be fitted to real data with simple novel methodology.

We think that all these properties make these constructions appealing in several applications.

Acknowledgments The first author acknowledges the support by Faculty of Economics and Management at Free University of Bozen-Bolzano via the project MUD.

References

1. Amblard, C., Girard, S.: A new extension of bivariate FGM copulas. Metrika **70**, 1–17 (2009)
2. Bernard, C., Liu, Y., MacGillivray, N., Zhang, J.: Bounds on capital requirements for bivariate risk with given marginals and partial information on the dependence. Depend. Model. **1**, 37–53 (2013)
3. Cuadras, C.M., Augé, J.: A continuous general multivariate distribution and its properties. Commun. Stat. A-Theory Methods **10**(4), 339–353 (1981)
4. Durante, F.: Generalized composition of binary aggregation operators. Int. J. Uncertain. Fuzziness Knowl.-Based Syst. **13**(6), 567–577 (2005)
5. Durante, F.: A new class of symmetric bivariate copulas. J. Nonparametr. Stat. **18**(7–8), 499–510 (2006/2007)
6. Durante, F., Fernández-Sánchez, J., Pappadà, R.: Copulas, diagonals and tail dependence. Fuzzy Sets Syst. **264**, 22–41 (2015)
7. Durante, F., Foschi, R., Spizzichino, F.: Threshold copulas and positive dependence. Stat. Probab. Lett. **78**(17), 2902–2909 (2008)
8. Durante, F., Kolesárová, A., Mesiar, R., Sempi, C.: Semilinear copulas. Fuzzy Sets Syst. **159**(1), 63–76 (2008)
9. Durante, F., Okhrin, O.: Estimation procedures for exchangeable Marshall copulas with hydrological application. Stoch. Environ. Res Risk Assess **29**(1), 205–226 (2015)
10. Durante, F., Quesada-Molina, J.J., Úbeda-Flores, M.: On a family of multivariate copulas for aggregation processes. Inf. Sci. **177**(24), 5715–5724 (2007)
11. Durante, F., Salvadori, G.: On the construction of multivariate extreme value models via copulas. Environmetrics **21**(2), 143–161 (2010)
12. Durante, F., Sempi, C.: On the characterization of a class of binary operations on bivariate distribution functions. Publ. Math. Debrecen **69**(1–2), 47–63 (2006)
13. Frahm, G.: On the extremal dependence coefficient of multivariate distributions. Stat. Probab. Lett. **76**(14), 1470–1481 (2006)
14. Genest, C., Carabarím-Aguirre, A., Harvey, F.: Copula parameter estimation using Blomqvist's beta. J. SFdS **154**(1), 5–24 (2013)
15. Genest, C., Favre, A.C.: Everything you always wanted to know about copula modeling but were afraid to ask. J. Hydrol. Eng. **12**(4), 347–368 (2007)
16. Joe, H.: Dependence Modeling with Copulas. Chapman and Hall/CRC, London (2014)
17. Kaas, R., Laeven, R.J.A., Nelsen, R.B.: Worst VaR scenarios with given marginals and measures of association. Insur. Math. Econom. **44**(2), 146–158 (2009)
18. Krupskii, P., Joe, H.: Factor copula models for multivariate data. J. Multivar. Anal. **120**, 85–101 (2013)
19. Li, X., Pellerey, F.: Generalized Marshall–Olkin distributions and related bivariate aging properties. J. Multivar. Anal. **102**(10), 1399–1409 (2011)
20. Liebscher, E.: Construction of asymmetric multivariate copulas. J. Multivar. Anal. **99**(10), 2234–2250 (2008)

21. Mai, J.F., Schenk, S., Scherer, M.: Exchangeable exogenous shock models. Bernoulli (2015). In press
22. Mai, J.F., Scherer, M.: Lévy-Frailty copulas. J. Multivar. Anal. **100**(7), 1567–1585 (2009)
23. Marshall, A.W.: Copulas, marginals, and joint distributions. In: Distributions with Fixed Marginals and Related Topics (Seattle, WA, 1993). IMS Lecture Notes Monograph Series, vol. 28, pp. 213–222. Institute of Mathematical Statistics, Hayward, CA (1996)
24. Marshall, A.W., Olkin, I.: A multivariate exponential distribution. J. Am. Stat. Assoc. **62**, 30–44 (1967)
25. Marshall, A.W., Olkin, I., Arnold, B.C.: Inequalities: theory of majorization and its applications, 2nd edn. Springer Series in Statistics. Springer, New York (2011)
26. Mazo, G., Girard, S., Forbes, F.: A flexible and tractable class of one-factor copulas. Statistics and Computing (2015), to appear. doi:10.1007/s11222-015-9580-7
27. Mazo, G., Girard, S., Forbes, F.: Weighted least-squares inference based on dependence coefficients for multivariate copulas (2014). http://hal.archives-ouvertes.fr/hal-00979151
28. Muliere, P., Scarsini, M.: Characterization of a Marshall–Olkin type class of distributions. Ann. Inst. Stat. Math. **39**(2), 429–441 (1987)
29. Pinto, J., Kolev, N.: Extended Marshall–Olkin model and its dual version. In: Cherubini, U., Durante, F., Mulinacci, S. (eds.) Marshall–Olkin Distributions—Advances in Theory and Applications, Springer Proceedings in Mathematics & Statistics. Springer International Publishing, Switzerland (2015)
30. Salvadori, G., De Michele, C., Durante, F.: On the return period and design in a multivariate framework. Hydrol. Earth Syst. Sci. **15**, 3293–3305 (2011)
31. Salvadori, G., Durante, F., De Michele, C.: Multivariate return period calculation via survival functions. Water Resour. Res. **49**(4), 2308–2311 (2013)
32. Sklar, A.: Fonctions de répartition à n dimensions et leurs marges. Publ. Inst. Stat. Univ. Paris **8**, 229–231 (1959)
33. Tankov, P.: Improved Fréchet bounds and model-free pricing of multi-asset options. J. Appl. Probab. **48**(2), 389–403 (2011)

Chapter 3
The Mean of Marshall–Olkin-Dependent Exponential Random Variables

Lexuri Fernández, Jan-Frederik Mai and Matthias Scherer

Abstract The probability distribution of $S_d := X_1 + \cdots + X_d$, where the vector $(X_1, \ldots, X_d)$ is distributed according to the Marshall–Olkin law, is investigated. Closed-form solutions are derived in the general bivariate case and for $d \in \{2, 3, 4\}$ in the exchangeable subfamily. Our computations can, in principle, be extended to higher dimensions, which, however, becomes cumbersome due to the large number of involved parameters. For the Marshall–Olkin distributions with conditionally independent and identically distributed components, however, the limiting distribution of S_d/d is identified as d tends to infinity. This result might serve as a convenient approximation in high-dimensional situations. Possible fields of application for the presented results are reliability theory, insurance, and credit-risk modeling.

3.1 Introduction

The distribution of $S_d := X_1 + \cdots + X_d$ has been treated considerably in the literature. For mathematical tractability, the individual random variables X_k are often considered to be independent, see, e.g., [2], a hypothesis that is hardly ever met in real-world applications. Another case where the distribution of the sum is known is when $(X_1, \ldots, X_d)$ has an elliptical distribution, see [10], a stability result that (at least partially) explains the popularity of elliptical distributions. Again, it could be

Lexuri Fernandez is grateful to UPV/EHU for the FPI-UPV/EHU grant.

L. Fernández (✉) · M. Scherer
Lehrstuhl für Finanzmathematik (M13), Technische Universität München, Parkring 11, 85748 Garching-Hochbrück, Germany
e-mail: lexuri.fernandez@tum.de

M. Scherer
e-mail: scherer@tum.de

J.-F. Mai
XAIA Investment, Sonnenstraße 19, 80331 München, Germany
e-mail: jan-frederik.mai@xaia.com

U. Cherubini et al. (eds.), *Marshall–Olkin Distributions - Advances in Theory and Applications*, Springer Proceedings in Mathematics & Statistics 141,
DOI 10.1007/978-3-319-19039-6_3

that this distributional assumption does not hold for the application one has in mind. In our study, we assume $(X_1, \ldots, X_d)$ to be distributed according to the Marshall–Olkin law; a popular assumption for dependent lifetimes in insurance and credit-risk modeling, see [9, 13]. With this interpretation in mind, S_d/d denotes the average lifetime of dependent exponential random variables. Applications might be the costs of an insurance company in the case of a natural catastrophe ([7]) or maintenance fees that have to be paid as long as some system is working. Related studies on the probability law of a sum of dependent risks can be found in the literature related to insurance and risk-management, see, e.g., ([1, 5, 22, 25]).

We derive $\mathbb{P}(S_2 > x)$ explicitly in the general case. One property of the Marshall–Olkin law is the large number of parameters, namely $2^d - 1$ in dimension d, rendering the Marshall–Olkin law challenging to work with as d increases. To account for this, in Sect. 3.3, we focus on the exchangeable subfamily, which has only d parameters in dimension d. In our case, we compute $\mathbb{P}(S_d > x)$ for $d \in \{2, 3, 4\}$. We guide the interested reader to strategies how extensions to higher dimensions might be achieved. Moreover, we study the asymptotic distribution of S_d/d (when $d \to \infty$) in the subfamily of Marshall–Olkin distributions with conditionally i.i.d. (CIID) components. Upper bounds for the sum of exchangeable vectors of CIID variables are already studied, see [6]. In [25] the asymptotic quantile behavior of a sum of dependent variables, where the dependence structure is given by an Archimedean copula, is analyzed. In our situation, the limiting case is related to certain exponential functionals of Lévy subordinators which are studied, e.g., in [3, 14, 16, 23].

The paper is organized as follows: In Sect. 3.2 the general Marshall–Olkin distribution is introduced and we compute the distribution of S_2. Section 3.3 considers the exchangeable case and computes $\mathbb{P}(S_d > x)$ for $d \in \{2, 3, 4\}$. In Sect. 3.4 we analyze the asymptotic case $d \to \infty$. Section 3.5 concludes.

3.2 The Marshall–Olkin Law

Marshall and Olkin [21] introduce a d-dimensional exponential distribution by lifting the univariate *lack of memory property* $\mathbb{P}(X > x + y | X > y) = \mathbb{P}(X > x)$, for all $x, y > 0$, to higher dimensions. If X is supported on $[0, \infty)$ and satisfies the univariate *lack of memory property*, then X is exponentially distributed. If $(X_1, \ldots, X_d)$ and all possible subvectors $(X_{i_1}, \ldots, X_{i_k})$, where $1 \leq i_1 < \cdots < i_k \leq d$, satisfy the multidimensional *lack of memory property*,

$$\begin{aligned}\mathbb{P}(X_{i_1} > x_{i_1} + y, \ldots, X_{i_k} > x_{i_k} + y \mid X_{i_1} > y, \ldots, X_{i_k} > y) \\ = \mathbb{P}(X_{i_1} > x_{i_1}, \ldots, X_{i_k} > x_{i_k}),\end{aligned} \tag{3.1}$$

where $x_{i_1}, \ldots, x_{i_k}, y > 0$, it is shown in [21] that the only distribution with support $[0, \infty)^d$ satisfying condition (3.1) is characterized by the survival function introduced in Definition 3.1 below.

Definition 3.1 (*Marshall–Olkin distribution*) Let $(X_1, \dots, X_d)$ represent a system of residual lifetimes with support $[0, \infty)^d$. Assume that the remaining components in this vector have a joint distribution that is independent of the age of the system, i.e., $(X_1, \dots, X_d)$ satisfies the multidimensional *lack of memory property* (3.1). Then, for $x_1, \dots, x_d \geq 0$,

$$\bar{F}(x_1, \dots, x_d) := \mathbb{P}(X_1 > x_1, \dots, X_d > x_d) = \exp\left(- \sum_{\emptyset \neq I \subseteq \{1,\dots,d\}} \lambda_I \max_{i \in I}\{x_i\} \right) \quad (3.2)$$

for certain parameters $\lambda_I \geq 0$, $\emptyset \neq I \subseteq \{1, \dots, d\}$, and $\sum_{I:k \in I} \lambda_I > 0$, $k = 1, \dots, d$. This multivariate probability law is called Marshall–Olkin distribution.

This distribution has key impact in reliability theory [8, 21], credit-risk management [13], and insurance [9]. Interpreting X_k as lifetime of component k, λ_I represents the intensity of the arrival time of a "shock" influencing the lifetime of all components in I. This can be seen from the canonical construction of the Marshall–Olkin distribution which is the following *fatal-shock* model, see [8, 20]. Let E_I, $\emptyset \neq I \subseteq \{1, \dots, d\}$, be exponentially distributed random variables with parameters $\lambda_I \geq 0$. We assume all E_I to be independent and interpret them as the arrival times of exogenous shocks to the respective components in I and define

$$X_k := \min\{E_I | \emptyset \neq I \subseteq \{1, \dots, d\}, k \in I\} \in (0, \infty), \quad k = 1, \dots, d, \quad (3.3)$$

where the variable X_k is the first time a shock hits component[1] k. The random vector $(X_1, \dots, X_d)$ as defined in Eq. (3.3) follows the Marshall–Olkin distribution.

Next, we derive the probability distribution of $S_2 = aX_1 + bX_2$, where a, b are positive constants. Providing an interpretation, with $a = b = 1/2$ the quantity $S_2/2$ is precisely the average lifetime of the two components. To simplify notation, we write $\lambda_1, \lambda_2, \lambda_{12}$ instead of $\lambda_{\{1\}}, \lambda_{\{2\}}, \lambda_{\{1,2\}}$ and we write E_1, E_2, E_{12} instead of $E_{\{1\}}, E_{\{2\}}, E_{\{1,2\}}$.

Lemma 3.1 (The weighted sum of two lifetimes) *On the probability space $(\Omega, \mathcal{F}, \mathbb{P})$ let (X_1, X_2) be a random vector constructed as in (3.3) and a, b positive constants. The survival function of the weighted sum of X_1 and X_2 is computed as*

$$\begin{aligned} &\mathbb{P}(aX_1 + bX_2 > x) \\ &= \frac{\lambda_1}{\lambda_1 - (\lambda_2 + \lambda_{12})\frac{a}{b}} e^{-(\lambda_2+\lambda_{12})\frac{x}{b}} \left(1 - e^{-(\lambda_1 - (\lambda_2+\lambda_{12})\frac{a}{b})\frac{x}{a+b}}\right) \\ &\quad + \frac{\lambda_2}{\lambda_2 - (\lambda_1 + \lambda_{12})\frac{b}{a}} e^{-(\lambda_1+\lambda_{12})\frac{x}{a}} \left(1 - e^{-(\lambda_2 - (\lambda_1+\lambda_{12})\frac{b}{a})\frac{x}{a+b}}\right) \\ &\quad + e^{-(\lambda_1+\lambda_2+\lambda_{12})\frac{x}{a+b}}. \end{aligned} \quad (3.4)$$

[1] The parameters $\lambda_I \geq 0$ represent the intensities of the exogenous shocks. Some of these can be 0, in which case $E_I \equiv \infty$. We require $\sum_{\emptyset \neq I:k \in I} \lambda_I > 0$, so for each $k = 1, \dots, d$ there is at least one subset $I \subseteq \{1, \dots, d\}$, containing k, such that $\lambda_I > 0$. Therefore, (3.3) is well-defined.

Proof

$$\mathbb{P}(aX_1+bX_2>x)=\mathbb{P}(aX_1+bX_2>x, X_1<X_2)+\mathbb{P}(aX_1+bX_2>x, X_2<X_1)$$
$$+\mathbb{P}(aX_1+bX_2>x, X_1=X_2).$$

Observe that, $X_1 < X_2 \Leftrightarrow E_1 < X_2$, $X_2 < X_1 \Leftrightarrow E_2 < X_1$, $X_1 = X_2 \Leftrightarrow E_{12} < \min\{E_1, E_2\}$, and, $\min\{E_1, E_2\} \sim \mathrm{Exp}(\lambda_1+\lambda_2)$.

Then,

$$\begin{aligned}
&\mathbb{P}(aX_1+bX_2>x, X_1<X_2)\\
&=\mathbb{P}(aX_1+bX_2>x, E_1<X_2)=\mathbb{P}\left(X_2>E_1>\frac{x-bX_2}{a}\right)\\
&=\mathbb{E}\left[\mathbb{P}\left(X_2>E_1>\frac{x-bX_2}{a}|E_1\right)\right]=\int_0^\infty \mathbb{P}\left(X_2>y_1>\frac{x-bX_2}{a}\right)f_{E_1}(y_1)dy_1\\
&=\frac{\lambda_1}{\lambda_1-(\lambda_2+\lambda_{12})\frac{a}{b}}e^{-(\lambda_2+\lambda_{12})\frac{x}{b}}\left(1-e^{-(\lambda_1-(\lambda_2+\lambda_{12})\frac{a}{b})\frac{x}{a+b}}\right)\\
&+\frac{\lambda_1}{\lambda_1+\lambda_2+\lambda_{12}}e^{-(\lambda_1+\lambda_2+\lambda_{12})\frac{x}{a+b}}.
\end{aligned}$$

$\mathbb{P}(aX_1+bX_2>x, X_2<X_1)$ and $\mathbb{P}(aX_1+bX_2>x, X_1=X_2)$ are computed in the same way.

Figure 3.1 plots the survival function (above) and the density function of $S_2 = aX_1+bX_2$, where $a = 30\,\%$ and $b = 70\,\%$.

Once the survival function of S_2 is known, one can further compute the density and the Laplace transform of S_2.

Corollary 3.1 *Let $(\Omega, \mathscr{F}, \mathbb{P})$ be a probability space, (X_1, X_2) a random vector constructed as in (3.3), and a, b positive constants. Then the Laplace transform of $S_2 = aX_1+bX_2$ is given by*

$$\begin{aligned}
\psi_{S_2}(t)&=\mathbb{E}\left[e^{-tS_2}\right]\\
&=\frac{\lambda_1(\lambda_2+\lambda_{12})b}{(\lambda_1 b-(\lambda_2+\lambda_{12})a)(\lambda_2+\lambda_{12}+tb)}+\frac{\lambda_2(\lambda_1+\lambda_{12})a}{(\lambda_2 a-(\lambda_1+\lambda_{12})b)(\lambda_1+\lambda_{12}+ta)}\\
&\quad-\left(\frac{\lambda_1(\lambda_2+\lambda_{12})}{\lambda_1 b-(\lambda_2+\lambda_{12})a}+\frac{\lambda_2(\lambda_1+\lambda_{12})}{\lambda_2 a-(\lambda_1+\lambda_{12})b}\right)\frac{a+b}{\lambda_1+\lambda_2+\lambda_{12}+t(a+b)}\\
&\quad+\frac{\lambda_{12}}{\lambda_1+\lambda_2+\lambda_{12}+t(a+b)}.
\end{aligned} \tag{3.5}$$

Proof We first need to compute the probability density function:

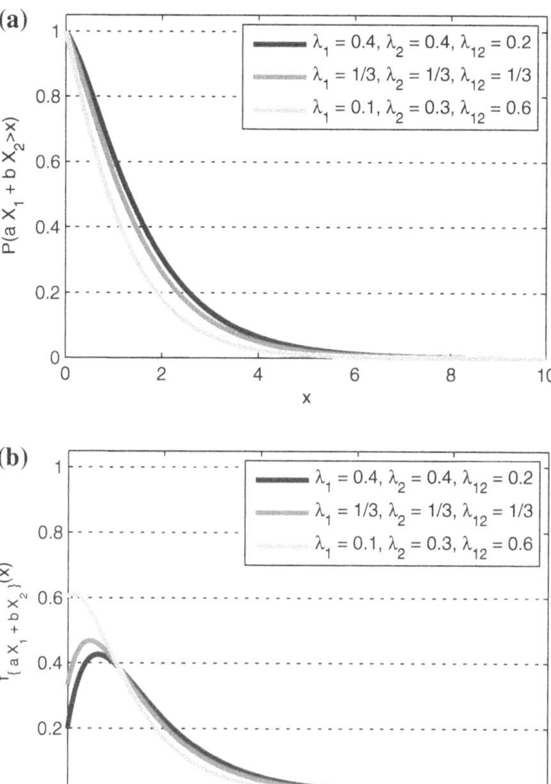

Fig. 3.1 The survival and density function of $S_2 = aX_1 + bX_2$, where $a = 30\,\%$ and $b = 70\,\%$

$$
\begin{aligned}
f_{S_2}(x) &= \frac{d}{dx}\left(1 - F_{S_2}(x)\right) \\
&= \frac{\lambda_1(\lambda_2+\lambda_{12})}{\lambda_1 b - (\lambda_2+\lambda_{12})a} e^{-(\lambda_2+\lambda_{12})x/b}\left(1 - e^{-(\lambda_1-(\lambda_2+\lambda_{12})a/b)\frac{x}{a+b}}\right) \\
&\quad + \frac{\lambda_2(\lambda_1+\lambda_{12})}{\lambda_2 a - (\lambda_1+\lambda_{12})b} e^{-(\lambda_1+\lambda_{12})x/a}\left(1 - e^{-(\lambda_2-(\lambda_1+\lambda_{12})b/a)\frac{x}{a+b}}\right) \\
&\quad + \frac{\lambda_{12}}{a+b} e^{-(\lambda_1+\lambda_2+\lambda_{12})\frac{x}{a+b}}.
\end{aligned} \tag{3.6}
$$

So, the Laplace transform is computed by evaluating the integral

$$\psi_{S_2}(t) = \int_0^\infty e^{-tx} f_{S_2}(x)dx.$$

Remark 3.1 Note that when $\lambda_1 - (\lambda_2 + \lambda_{12})a/b = 0$ or $\lambda_2 - (\lambda_1 + \lambda_{12})b/a = 0$, Eqs. (3.4), (3.5), and (3.6) are not defined. By computing the respective limits (that do exist!) when the parameters approach such a constellation, the functions can be extended continuously.

If one aims at generalizing these results to higher dimensions, one notices that the number of involved shocks and parameters, i.e., $2^d - 1$ in dimension d, renders this problem extremely intractable already for moderate dimensions d. A subclass with fewer parameters is obtained by considering the Marshall–Olkin law with exchangeable components. This yields a parametric family with d parameters in dimension d, allowing us to derive the distribution of S_d in higher dimensions.

3.3 The Exchangeable Marshall–Olkin Law

The aim of this section is to compute the survival function of S_d in the exchangeable case. We introduce the subfamily of exchangeable Marshall–Olkin laws in order to deal with the problem of overparameterization. For a deeper background on exchangeable Marshall–Olkin laws see [18, 19] (Chap. 3, Sect. 3.2). A random vector $(X_1, \ldots, X_d)$ is said to be exchangeable if for all permutations π on $\{1, \ldots, d\}$ it satisfies

$$\mathbb{P}(X_1 > x_1, \ldots, X_d > x_d) = \mathbb{P}(X_1 > x_{\pi(1)}, \ldots, X_d > x_{\pi(d)}), \quad x_1, \ldots, x_d \in \mathbb{R}, \tag{3.7}$$

or, alternatively in the Marshall–Olkin context, if the exchangeability condition

$$|I| = |J| \Rightarrow \lambda_I = \lambda_J, \tag{3.8}$$

is met. The proof that (3.8) is equivalent to $(X_1, \ldots, X_d)$ being exchangeable can be found in [19], page 124. Condition (3.8) means that two shocks affecting subsets with identical cardinalities have the same intensity λ_I. Hence, in this section we denote by λ_1 the intensity of all shocks affecting precisely one component, by λ_2 all shocks affecting two components, and so on.

Let $(X_1, \ldots, X_d)$ be a random vector following the Marshall–Olkin distribution, defined as in Eq. (3.3). Then the survival function of the exchangeable Marshall–Olkin law is given, for $x_1, \ldots, x_d \geq 0$, by

$$\bar{F}(x_1, \ldots, x_d) = \exp\left(-\sum_{k=1}^{d} x_{(d+1-k)} \sum_{i=0}^{d-k} \binom{d-k}{i} \lambda_{i+1} \right), \tag{3.9}$$

$x_{(1)} \leq \cdots \leq x_{(d)}$ being the ordered list of $x_1, \ldots, x_d$.

Observe that now instead of dealing with $2^d - 1$ parameters λ_I we just have to work with d parameters $\lambda_1, \ldots, \lambda_d$, which simplifies the process of computing the required probabilities.

In the following, we present the survival function of the sum of components of Marshall–Olkin random vectors in low dimensional exchangeable cases (2, 3, and 4-dimensional).

Lemma 3.2 (The sum of $d \in \{2, 3, 4\}$ lifetimes) *On the probability space* $(\Omega, \mathscr{F}, \mathbb{P})$ ...

1. ... *let* (X_1, X_2) *be a two-dimensional exchangeable Marshall–Olkin random vector. Then, for* $x \geq 0$,

$$\mathbb{P}(X_1 + X_2 > x) = \frac{2\lambda_1 e^{-(\lambda_1+\lambda_2)x}}{\lambda_2}\left(e^{\lambda_2 \frac{x}{2}} - 1\right) + e^{-(2\lambda_1+\lambda_2)\frac{x}{2}}. \quad (3.10)$$

2. ... *let* (X_1, X_2, X_3) *be a three-dimensional exchangeable Marshall–Olkin random vector. Then, for* $x \geq 0$,

$$\begin{aligned}\mathbb{P}(X_1 + X_2 + X_3 > x) = e^{-(3\lambda_1+3\lambda_2+\lambda_3)\frac{x}{3}} &+ \frac{6\lambda_1(2\lambda_1 + 3\lambda_2 + \lambda_3)}{(3\lambda_2 + \lambda_3)(\lambda_2 + \lambda_3)} \times \\ &e^{-(2\lambda_1+3\lambda_2+\lambda_3)\frac{x}{2}}\left(e^{\left(\frac{3\lambda_2+\lambda_3}{2}\right)\frac{x}{3}} - 1\right) \\ &+ \frac{3\lambda_2(\lambda_2 + \lambda_3) - 6\lambda_1(\lambda_1 + \lambda_2)}{(\lambda_2 + \lambda_3)(3\lambda_2 + 2\lambda_3)} \times \\ &e^{-(\lambda_1+2\lambda_2+\lambda_3)x}\left(e^{(3\lambda_2+2\lambda_3)\frac{x}{3}} - 1\right).\end{aligned} \quad (3.11)$$

3. ... *let* (X_1, X_2, X_3, X_4) *be a four-dimensional exchangeable Marshall–Olkin random vector. Then, for* $x \geq 0$,

$$\begin{aligned}\mathbb{P}(X_1 + X_2 + X_3 + X_4 > x) &= 24 \cdot P_1 + 12 \cdot P_2 + 12 \cdot P_3 \\ &+ 12 \cdot P_4 + 4 \cdot P_5 + 4 \cdot P_6 + 6 \cdot P_7 + P_8,\end{aligned} \quad (3.12)$$

where,

$$\begin{aligned}P_1 &= \mathbb{P}(X_1 + X_2 + X_3 + X_4 > x | X_1 < X_2 < X_3 < X_4) \\ &= \lambda_1(\lambda_1 + \lambda_2) f_{11}\Big(\frac{32 f_{10}}{f_1 f_2 f_4 f_5} e^{-f_1 \frac{x}{4}} - \frac{27 f_{10}}{f_2 f_3 f_7 f_8} e^{-f_3 \frac{x}{3}} + \frac{4 f_{10}}{f_4 f_6 f_7 f_9} e^{-f_9 \frac{x}{2}} \\ &\quad - \frac{1}{f_5 f_6 f_8} e^{-f_{10} x}\Big), \\ P_2 &= \mathbb{P}(X_1 + X_2 + X_3 + X_4 > x | X_1 < X_2 < X_3 = X_4) \\ &= \lambda_1(\lambda_1 + \lambda_2) f_6\Big(\frac{8}{f_1 f_2 f_4} e^{-f_1 \frac{x}{4}} - \frac{9}{f_2 f_3 f_7} e^{-f_3 \frac{x}{3}} + \frac{2}{f_4 f_7 f_9} e^{-f_9 \frac{x}{2}}\Big),\end{aligned}$$

$$
\begin{aligned}
P_3 &= \mathbb{P}(X_1 + X_2 + X_3 + X_4 > x | X_1 < X_2 = X_3 < X_4) \\
&= \lambda_1(\lambda_2 + \lambda_3)\Big[\frac{f_{10}}{f_2}\Big(\frac{16}{f_1 f_5}e^{-f_1\frac{x}{4}} - \frac{9}{f_3 f_8}e^{-f_3\frac{x}{3}}\Big) + \frac{1}{f_5 f_8}e^{-f_{10}x}\Big], \\
P_4 &= \mathbb{P}(X_1 + X_2 + X_3 + X_4 > x | X_1 = X_2 < X_3 < X_4) \\
&= \lambda_2 f_{11}\Big[\Big(\frac{2f_{10}}{f_4 f_6 f_9} - \frac{1}{f_5 f_6} + \frac{1}{f_1 f_9}\Big)e^{-f_1\frac{x}{4}} - \frac{2f_{10}}{f_4 f_6 f_9}e^{-f_9\frac{x}{2}} + \frac{1}{f_5 f_6}e^{-f_{10}x}\Big], \\
P_5 &= \mathbb{P}(X_1 + X_2 + X_3 + X_4 > x | X_1 = X_2 = X_3 < X_4) \\
&= \lambda_3\Big(\frac{4f_{10}}{f_1 f_5}e^{-f_1\frac{x}{4}} - \frac{1}{f_5}e^{-f_{10}x}\Big), \\
P_6 &= \mathbb{P}(X_1 + X_2 + X_3 + X_4 > x | X_1 < X_2 = X_3 = X_4) \\
&= \frac{\lambda_1}{f_2}(\lambda_3 + \lambda_4)\Big(\frac{4}{f_1}e^{-f_1\frac{x}{4}} - \frac{3}{f_3}e^{-f_3\frac{x}{3}}\Big),
\end{aligned}
$$

$$
\begin{aligned}
P_7 &= \mathbb{P}(X_1 + X_2 + X_3 + X_4 > x | X_1 = X_2 < X_3 = X_4) \\
&= \frac{\lambda_2 f_6}{f_4}\Big(\frac{2}{f_1}e^{-f_1\frac{x}{4}} - \frac{1}{f_9}e^{-f_9\frac{x}{2}}\Big), \\
P_8 &= \mathbb{P}(X_1 + X_2 + X_3 + X_4 > x | X_1 = X_2 = X_3 = X_4) \\
&= \frac{\lambda_4}{f_1}e^{-f_1\frac{x}{4}},
\end{aligned}
$$

and

$$
\begin{aligned}
f_1 &= 4\lambda_1 + 6\lambda_2 + 4\lambda_3 + \lambda_4, & f_5 &= 6\lambda_2 + 8\lambda_3 + 3\lambda_4, & f_9 &= 2\lambda_1 + 5\lambda_2 + 4\lambda_3 + \lambda_4, \\
f_2 &= 6\lambda_2 + 4\lambda_3 + \lambda_4, & f_6 &= \lambda_2 + 2\lambda_3 + \lambda_4, & f_{10} &= \lambda_1 + 3\lambda_2 + 3\lambda_3 + \lambda_4, \\
f_3 &= 3\lambda_1 + 6\lambda_2 + 4\lambda_3 + \lambda_4, & f_7 &= 3\lambda_2 + 4\lambda_3 + \lambda_4, & f_{11} &= \lambda_1 + 2\lambda_2 + \lambda_3. \\
f_4 &= 4\lambda_2 + 4\lambda_3 + \lambda_4, & f_8 &= 3\lambda_2 + 5\lambda_3 + 2\lambda_4, & &
\end{aligned}
$$

Proof We prove the case $d = 2$, considering that the proofs for $d = 3$ and $d = 4$ are done in the same way.

$$
\begin{aligned}
\mathbb{P}(X_1 + X_2 > x) = 2\mathbb{P}(X_1 + X_2 > x | X_1 < X_2)\mathbb{P}(X_1 < X_2) \\
+\mathbb{P}(X_1 + X_2 > x | X_1 = X_2)\mathbb{P}(X_1 = X_2).
\end{aligned}
$$

such that $E_1, E_2 \sim \text{Exp}(\lambda_1)$ and $E_{12} \sim \text{Exp}(\lambda_2)$ and note that since we are working on the exchangeable case,

$$\mathbb{P}(X_1 + X_2 > x | X_1 > X_2)\mathbb{P}(X_1 > X_2) = \mathbb{P}(X_1 + X_2 > x | X_1 < X_2)\mathbb{P}(X_1 < X_2).$$

Taking into account that, $X_1 < X_2 \Leftrightarrow E_1 < \min\{E_2, E_{12}\}$ and $X_1 = X_2 \Leftrightarrow \min\{E_1, E_2\} > E_{12}$,

$$\begin{aligned}\mathbb{P}(X_1+X_2>x) &= 2\mathbb{P}(E_1+\min\{E_2,E_{12}\}>x|E_1<\min\{E_2,E_{12}\})\mathbb{P}(E_1<\min\{E_2,E_{12}\})\\ &\quad+\mathbb{P}(E_{12}+E_{12}>x|E_{12}<\min\{E_1,E_2\})\mathbb{P}(E_{12}<\min\{E_1,E_2\})\\ &= 2\mathbb{E}\left[\mathbb{P}(\min\{E_2,E_{12}\}>E_1>x-\min\{E_2,E_{12}\}\,|E_1)\right]\\ &\quad+\mathbb{E}\Big[\mathbb{P}(\min\{E_1,E_2\}>E_{12}>\frac{x}{2}|\min\{E_1,E_2\})\Big].\end{aligned}$$

Then, from the so-called *min-stability* of the exponential distribution, $\min\{E_1,E_2\}\sim\text{Exp}(2\lambda_1)$ and $\min\{E_2,E_{12}\}\sim\text{Exp}(\lambda_1+\lambda_2)$,

$$\begin{aligned}&\mathbb{E}\left[\mathbb{P}(\min\{E_2,E_{12}\}>E_1>x-\min\{E_2,E_{12}\}\,|E_1)\right]\\ &=\frac{\lambda_1}{\lambda_2}e^{-(\lambda_1+\lambda_2)x}\left(e^{\lambda_2\frac{x}{2}}-1\right)+\frac{\lambda_1}{2\lambda_1+\lambda_2}e^{-(2\lambda_1+\lambda_2)\frac{x}{2}},\end{aligned}$$

$$\mathbb{E}\Big[\mathbb{P}(\min\{E_1,E_2\}>E_{12}>\frac{x}{2}|\min\{E_1,E_2\})\Big]=\frac{\lambda_2}{2\lambda_1+\lambda_2}e^{-(2\lambda_1+\lambda_2)\frac{x}{2}}.$$

So,

$$\begin{aligned}\mathbb{P}(X_1+X_2>x) &= 2\Big(\frac{\lambda_1}{\lambda_2}e^{-(\lambda_1+\lambda_2)x}\left(e^{\lambda_2\frac{x}{2}}-1\right)+\frac{\lambda_1}{2\lambda_1+\lambda_2}e^{-(2\lambda_1+\lambda_2)\frac{x}{2}}\Big)\\ &\quad+\frac{\lambda_2}{2\lambda_1+\lambda_2}e^{-(2\lambda_1+\lambda 2)\frac{x}{2}}\\ &= \frac{2\lambda_1e^{-(\lambda_1+\lambda_2)x}}{\lambda_2}\left(e^{\lambda_2\frac{x}{2}}-1\right)+e^{-(2\lambda_1+\lambda_2)\frac{x}{2}}.\end{aligned}$$

Note that (from Remark 3.2 below) in the case $d=3$:

$$\begin{aligned}\mathbb{P}(X_1+X_2+X_3>x) &= 6\mathbb{P}(X_1+X_2+X_3>x|X_1<X_2<X_3)\times\\ &\quad\mathbb{P}(X_1<X_2<X_3)\\ &\quad+3\mathbb{P}(X_1+X_2+X_3>x|X_1=X_2<X_3)\times\\ &\quad\mathbb{P}(X_1=X_2<X_3)\\ &\quad+3\mathbb{P}(X_1+X_2+X_3>x|X_1<X_2=X_3)\times\\ &\quad\mathbb{P}(X_1<X_2=X_3)\\ &\quad+\mathbb{P}(X_1+X_2+X_3>x|X_1=X_2=X_3)\times\\ &\quad\mathbb{P}(X_1=X_2=X_3)\end{aligned}$$

has to be computed and in $d=4$:

$$\begin{aligned}\mathbb{P}(X_1+X_2+X_3+X_4>x) &= 24\,\mathbb{P}(X_1+X_2+X_3+X_4>x|X_1<X_2<X_3<X_4)\times\\ &\quad\mathbb{P}(X_1<X_2<X_3<X_4)\\ &\quad+12\,\mathbb{P}(X_1+X_2+X_3+X_4>x|X_1<X_2<X_3=X_4)\times\end{aligned}$$

$$
\begin{aligned}
&\mathbb{P}(X_1 < X_2 < X_3 = X_4)\\
&+ 12\,\mathbb{P}(X_1 + X_2 + X_3 + X_4 > x | X_1 < X_2 = X_3 < X_4)\times\\
&\mathbb{P}(X_1 < X_2 = X_3 < X_4)\\
&+ 12\,\mathbb{P}(X_1 + X_2 + X_3 + X_4 > x | X_1 = X_2 < X_3 < X_4)\\
&\mathbb{P}(X_1 = X_2 < X_3 < X_4)\\
&+ 4\,\mathbb{P}(X_1 + X_2 + X_3 + X_4 > x | X_1 = X_2 = X_3 < X_4)\times\\
&\mathbb{P}(X_1 = X_2 = X_3 < X_4)\\
&+ 4\,\mathbb{P}(X_1 + X_2 + X_3 + X_4 > x | X_1 < X_2 = X_3 = X_4)\times\\
&\mathbb{P}(X_1 < X_2 = X_3 = X_4)\\
&+ 6\,\mathbb{P}(X_1 + X_2 + X_3 + X_4 > x | X_1 = X_2 < X_3 = X_4)\times\\
&\mathbb{P}(X_1 = X_2 < X_3 = X_4)\\
&+ \mathbb{P}(X_1 + X_2 + X_3 + X_4 > x | X_1 = X_2 = X_3 = X_4)\times\\
&\mathbb{P}(X_1 = X_2 = X_3 = X_4).
\end{aligned}
$$

Figure 3.2 illustrates the survival (above) and density (below) functions for S_d, $d = 2, 3, 4$, in the exchangeable case.

Remark 3.2 (Generalizing the results to higher dimensions) Marshall–Olkin multivariate distributions are not absolutely continuous, i.e., there is a positive probability that several components take the same value, $\mathbb{P}(X_1 = \cdots = X_d) > 0$. It is possible to compute the expression

$$\mathbb{P}(X_1 + \cdots + X_d > x, X_1 = \cdots = X_d), \tag{3.13}$$

for all dimensions $d \in \mathbb{N}$, by recalling Pascal's triangle.

$$
\begin{aligned}
PM_d^d &:= \mathbb{P}(X_1 + \cdots + X_d > x, X_1 = \cdots = X_d) \\
&= \frac{\lambda_d}{\sum_{i=0}^{d} \binom{d}{i}\lambda_i} e^{-\left(\sum_{i=0}^{d}\binom{d}{i}\lambda_i\right)\frac{x}{d}}, \quad \lambda_0 = 0.
\end{aligned} \tag{3.14}
$$

However, the generalization to arbitrary singular events is not that obvious. Observe that from a sum of d elements we have to take into account the cases where we have k equalities in the conditions of the conditional probabilities, $k \in \{0, \ldots, d-1\}$. The number of cases which have to be taken into account is given by the binomial coefficient $(d-1)$ *choose* (k).

Take for example the case $d = 4$:

1. Number of cases where $k = 0$, i.e., there is no equality in the condition: $\binom{3}{0} = 1$,

$$\mathbb{P}(X_1 + \cdots + X_4 > x, X_{i_1} < \cdots < X_{i_4}), \quad \text{where} \quad i_k \neq i_j \in \{1, 2, 3, 4\}.$$

2. Number of cases where there is one equality ($k = 1$) in the condition: $\binom{3}{1} = 3$,

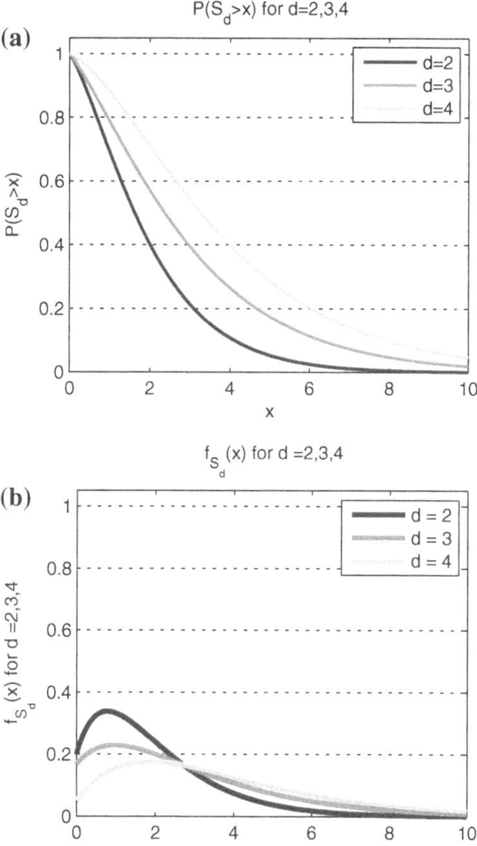

Fig. 3.2 Plots of the survival and density function for S_d, $d = 2, 3, 4$, in the exchangeable case. The parameters considered are in the two-dimensional case: $\lambda_1 = 0.6$, $\lambda_2 = 0.4$, in the three-dimensional case: $\lambda_1 = 0.1$, $\lambda_2 = 0.2$, $\lambda_3 = 0.5$, and when $d = 4$: $\lambda_1 = 0.05$, $\lambda_2 = 0.1$, $\lambda_3 = 0.15$, $\lambda_4 = 0.2$

$$\mathbb{P}(X_1 + \cdots + X_4 > x, X_{i_1} = X_{i_2} < X_{i_3} < X_{i_4}),$$
$$\mathbb{P}(X_1 + \cdots + X_4 > x, X_{i_1} < X_{i_2} = X_{i_3} < X_{i_4}),$$
$$\mathbb{P}(X_1 + \cdots + X_4 > x, X_{i_1} < X_{i_2} < X_{i_3} = X_{i_4}),$$

where $i_k \neq i_j \in \{1, 2, 3, 4\}$.

3. Number of cases where there are 2 equalities ($k = 2$) in the condition: $\binom{3}{2} = 3$,

$$\mathbb{P}(X_1 + \cdots + X_4 > x, X_{i_1} = X_{i_2} = X_{i_3} < X_{i_4}),$$
$$\mathbb{P}(X_1 + \cdots + X_4 > x, X_{i_1} < X_{i_2} = X_{i_3} = X_{i_4}),$$
$$\mathbb{P}(X_1 + \cdots + X_4 > x, X_{i_1} = X_{i_2} < X_{i_3} = X_{i_4}),$$

such that $i_k \neq i_j \in \{1, 2, 3, 4\}$.

Since we are in the exchangeable case, we need to calculate how many times each probability has to be added. For this purpose, let us consider the definition of *permutation of multisets*

$$PM_d^{a_1,a_2,\ldots,a_{k-1},a_k} := \frac{d!}{a_1! \cdot a_2! \cdot \ldots \cdot a_{k-1}! \cdot a_k!}, \tag{3.15}$$

where in our case $a_1, \ldots, a_k$ represent the numbers of elements which are equal and how they are located in each condition. Note that $\sum_{i=1}^{k} a_i = d$. Let us illustrate this relation with the example of $d = 4$:

$$\begin{aligned}
&\mathbb{P}(X_1 + \cdots + X_4 > x) \\
&= PM_4^{1,1,1,1} \cdot \mathbb{P}(X_1 + \cdots + X_4 > x, \underbrace{X_1}_{1} < \underbrace{X_2}_{1} < \underbrace{X_3}_{1} < \underbrace{X_4}_{1}) \\
&\quad + PM_4^{2,1,1} \cdot \mathbb{P}(X_1 + \cdots + X_4 > x, \underbrace{X_1 = X_2}_{2} < \underbrace{X_3}_{1} < \underbrace{X_4}_{1}) \\
&\quad + PM_4^{1,2,1} \cdot \mathbb{P}(X_1 + \cdots + X_4 > x, \underbrace{X_1}_{1} < \underbrace{X_2 = X_3}_{2} < \underbrace{X_4}_{1}) \\
&\quad + PM_4^{1,1,2} \cdot \mathbb{P}(X_1 + \cdots + X_4 > x, \underbrace{X_1}_{1} < \underbrace{X_2}_{1} < \underbrace{X_3 = X_4}_{2}) \\
&\quad + PM_4^{2,2} \cdot \mathbb{P}(X_1 + \cdots + X_4 > x, \underbrace{X_1 = X_2}_{2} < \underbrace{X_3 = X_4}_{2}) \\
&\quad + PM_4^{3,1} \cdot \mathbb{P}(X_1 + \cdots + X_4 > x, \underbrace{X_1 = X_2 = X_3}_{3} < \underbrace{X_4}_{1}) \\
&\quad + PM_4^{1,3} \cdot \mathbb{P}(X_1 + \cdots + X_4 > x, \underbrace{X_1}_{1} < \underbrace{X_2 = X_3 = X_4}_{3}) + PM_4^4,
\end{aligned} \tag{3.16}$$

the expression for PM_4^4 is given in Eq. (3.14).

Example 3.1 (Illustrating the effect of different levels of dependence) In Fig. 3.3, examples for the survival and density function of S_4 for different levels of dependence are visualized.

(a) **Independence case**: Shocks arriving to just one element are the only ones present in the system, i.e., $\lambda_1 > 0$ and $\lambda_2 = \lambda_3 = \lambda_4 = 0$. In this case, the probability distribution of S_d follows the Erlang distribution with rate λ_1 and degrees of freedom 4.

(b) **Comonotonic case**: The shock arriving to all components at the same time is the only one influencing the system, i.e., $\lambda_1 = \lambda_2 = \lambda_3 = 0$ and $\lambda_4 > 0$, and the distribution of S_d is exponential with mean $4/\lambda_4$.

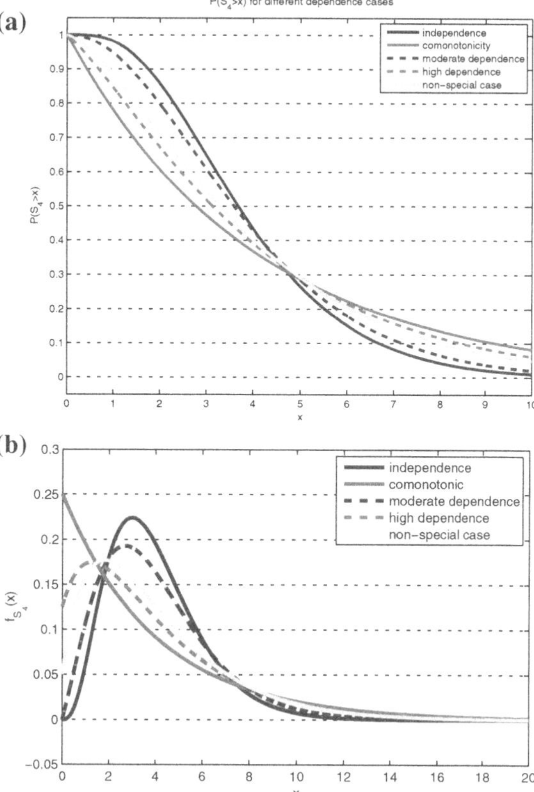

Fig. 3.3 $\mathbb{P}(S_4 > x)$ (*above*) and $f_{S_4}(x)$ (*below*) for different assumptions concerning the dependence: (a) independence, (b) comonotonicity, (c) moderate dependence (we consider $\lambda_4 = 0$), (d) high dependence (we consider $\lambda_1 = 0$), (e) non-special case. In all examples, the marginal laws are considered to be the same, X_i unit exponential random variables, $i = 1, \ldots, 4$

(c) **Moderate dependence case**: In this case, the shocks influencing fewer components jointly have the strongest influence, i.e., $\lambda_1 > \lambda_2 > \lambda_3 > \lambda_4 > 0$.
(d) **High dependence case**: Shocks arriving to most components jointly have the strongest influence, i.e., $\lambda_4 > \lambda_3 > \lambda_2 > \lambda_1 > 0$.
(e) **Non-special case**: $\lambda_1, \lambda_2, \lambda_3, \lambda_4 > 0$.

One can observe in Fig. 3.3 (above) that the intersection of the survival function is around the expected value $\mathbb{E}[S_4] = 4$. When the dependence between the components of the system is strong, the probability of the system to collapse before this intersection is lower than in the cases where the dependence is weak, but once the system survives till this intersection point, in cases with strong dependence the probability that the system will last alive longer is higher than in cases where the dependence is weak. This interpretation can be also seen in the densities (see Fig. 3.3, below). In weak dependence cases, the mass of the probability is concentrated around the expected value, which is translated into exhibiting a very steep decline of the survival function (see Fig. 3.3).

3.4 The Extendible Marshall–Olkin Law

In this section, we show how the probability distribution of S_d/d behaves in the limit when the system grows in dimension, i.e., for $d \to \infty$. For this purpose we work with the extendible subfamily of the Marshall–Olkin law, since we must be able to extend the dimension of the vector $(X_1, \ldots, X_d)$ without destroying its distributional structure. Recall that a random vector is called extendible if there exists an infinite exchangeable sequence $\{\tilde{X}_k\}_{k\in\mathbb{N}}$ such that $(X_1, \ldots, X_d) \stackrel{\mathscr{L}}{=} (\tilde{X}_1, \ldots, \tilde{X}_d)$. De Finetti's Theorem states that this is equivalent to $(\tilde{X}_1, \ldots, \tilde{X}_d)$ being conditionally i.i.d. (see [11]).

For extendible Marshall–Olkin laws there is a canonical construction based on Lévy subordinators, which are non-decreasing Lévy processes, $\{\Lambda_t, t \geq 0\}$, where the Lévy measure $\nu(dx)$ is defined on $\mathscr{B}((0, \infty])$ satisfying $\int (1 \wedge x)\nu(dx) < \infty$:

$$X_k = \inf\{t \geq 0 : \Lambda_t \geq \epsilon_k\}, \quad k = 1, \ldots, d. \tag{3.17}$$

Component X_k is the first-passage time of Λ across ϵ_k and $\{\epsilon_k\}_{k\in\mathbb{N}}$ is an i.i.d. sequence of unit exponential random variables. This construction is called the Lévy-frailty construction (for further information on these distributions we refer the reader to [17, 19]) and it defines the subclass of extendible Marshall–Olkin distributions.

Let $\{\Psi(k)\}_{k\in\mathbb{N}}$ be a sequence, derived from evaluating the Laplace exponent Ψ of Λ at the natural numbers. It is shown in [18] that

$$\mathbb{P}(X_1 > x_1, \ldots, X_d > x_d) = \exp\left(-\sum_{k=1}^{d} x_{(d-k+1)} \left(\Psi(k) - \Psi(k-1)\right) \right),$$

where $x_{(1)} \leq \cdots \leq x_{(d)}$ is the ordered list of the $x_1, \ldots, x_d \geq 0$ (see [17]), is the survival function of $(X_1, \ldots, X_d)$ which is completely determined by the sequence $\{\Psi(k)\}_{k\in\mathbb{N}}$. Then, $(X_1, \ldots, X_d)$ follows the Marshall–Olkin distribution with parameters

$$\lambda_k = \sum_{i=0}^{k-1} (-1)^i \left(\Psi(d-k+i+1) - \Psi(d-k+i)\right), \quad k = 1, \ldots, d.$$

Once we constructed the vector of first-passage times of a *Lévy subordinator*, $(X_1, \ldots, X_d)$, we can prove that when $d \to \infty$, S_d/d and the *exponential functional* of a Lévy subordinator, $I_\infty = \int_0^\infty e^{-\Lambda_s} ds$, have the same distribution. The exponential functional of a Lévy process, $\{\Lambda_t, t \geq 0\}$, is defined as

$$I_t = \int_0^t e^{-\Lambda_s} ds. \tag{3.18}$$

Lemma 3.3 *Let $(X_1, \ldots, X_d)$ be a random vector following the extendible Marshall–Olkin distribution. Then,*

$$\lim_{d \nearrow \infty} \frac{S_d}{d} \stackrel{\mathscr{L}}{=} I_\infty, \tag{3.19}$$

where $I_\infty = \int_0^\infty e^{-\Lambda_s} ds$ represents the exponential functional of the Lévy subordinator $\{\Lambda_t, t \geq 0\}$ at its terminal value. We refer the reader to [3] and [4] for detailed background on exponential functionals of Lévy processes.

Proof Define $X_k = \inf\{t \geq 0 : \Lambda_t \geq \epsilon_k\}$ as in Eq. (3.17). Then

$$\lim_{d \nearrow \infty} \frac{1}{d} \sum_{k=1}^{d} X_k \stackrel{\text{a.s.}}{=} \int_0^\infty e^{-\Lambda_s} ds \Leftrightarrow \mathbb{P}\left(\left| \lim_{d \nearrow \infty} \frac{1}{d} \sum_{k=1}^{d} X_k - \int_0^\infty e^{-\Lambda_s} ds \right| = 0 \right) = 1.$$

The strong law of large numbers implies that $\lim_{d \nearrow \infty} \frac{1}{d} \sum_{k=1}^{d} X_k = \mathbb{E}[X_1 | \Lambda]$ holds almost surely.

Observe that,

$$\begin{aligned}
\mathbb{E}[X_1 | \Lambda] &= \int_0^\infty x d\mathbb{P}(X_1 \leq x | \Lambda) \\
&= \int_0^\infty x d\mathbb{P}(\epsilon_1 \leq \Lambda_x | \Lambda) \\
&= \int_0^\infty x d(1 - e^{-\Lambda_x}) \\
&= \int_0^\infty -x d(e^{-\Lambda_x}) \\
&= \Big[-x e^{-\Lambda_x} \Big]_{x=0}^{x=\infty} + \int_0^\infty e^{-\Lambda_x} dx \\
&= 0 + \int_0^\infty e^{-\Lambda_x} dx.
\end{aligned}$$

Remark that convergence *almost surely* implies convergence *in distribution.*

Example 3.2 (The limit of S_d/d in a Poisson-frailty model) We want to analyze the convergence of $\mathbb{P}(S_d/d > x)$, $d \geq 2$, $x \geq 0$, in the limit $d \to \infty$. Considering the standard Poisson process as an example, $N_t = \{N_t\}_{t \geq 0}$ with intensity $\beta > 0$, which is a Lévy subordinator. Bertoin and Yor [3] investigates the distribution of the exponential functional of a standard Poisson process,

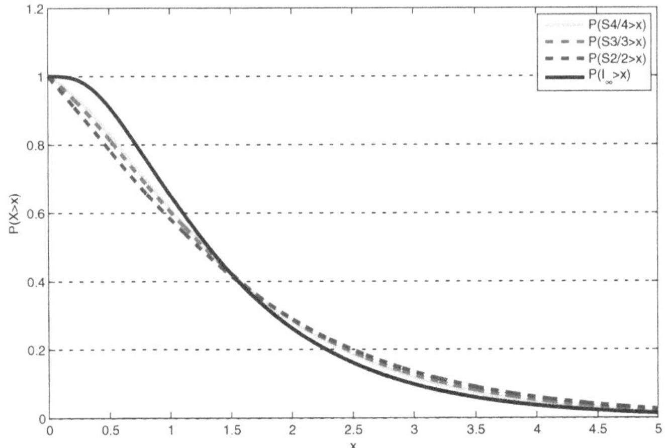

Fig. 3.4 Plot of $\mathbb{P}(S_d/d > x)$, $d = 2, 3, 4$ together with $\mathbb{P}(I_\infty > x)$, $x \geq 0$, where $\beta = 1$

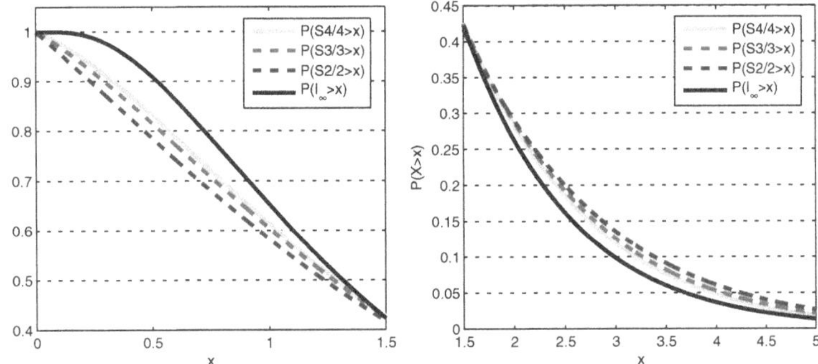

Fig. 3.5 Zoom into Fig. 3.4

$$I_\infty = \int_0^\infty e^{-N_t} dt, \tag{3.20}$$

using its Laplace transform:

$$\mathbb{E}[e^{\tilde{\lambda} I_\infty}] = \left(\prod_{j=0}^{\infty} (1 - \tilde{\lambda} e^{-j}) \right)^{-1}, \quad \tilde{\lambda} < 1. \tag{3.21}$$

Using the *Gaver–Stehfest* Laplace inversion technique (see [12, 15, 24]), we numerically compute the survival function of the exponential functional of I_∞ (Eq. (3.20)).

With this example we visualize how $\mathbb{P}(S_d/d > x)$, $d \in \mathbb{N}_0$, converges to $\mathbb{P}(I_\infty > x)$ when $d \to \infty$. In this case, the components of the system strongly depend on each other, i.e., $0 < \lambda_1 < \lambda_2 < \lambda_3 < \lambda_4$ (see Figs. 3.4 and 3.5).

3.5 Conclusion

We study the probability distribution of a sum of dependent random variables $S_d = X_1 + \cdots + X_d$ when the dependence structure is given by the Marshall–Olkin distribution. The Marshall–Olkin law possesses interesting properties from a statistical point of view as well as for applications in different fields like financial risk-management or insurance. However, during the construction of this type of dependence structure, we encounter the obstacle of overparameterization. In order to deal with this drawback and to make the computations more tractable, we work with the exchangeable subfamily, where the amount of parameters is significantly decreased from $2^d - 1$ to d. In low dimensional cases, $d = 2$, $d = 3$, and $d = 4$, we develop the explicit expressions for the distribution of S_d and we give a sketch of how these results can be extended to higher dimensions.

However, note that while the number of factors in the sum increases in one unit, the number of cases into consideration for the calculus of the probabilities increases in 2^{d-1}. This is the reason why the problem becomes intractable for $d > 4$ and we focus on analyzing the behavior of S_d/d in the limiting case, $d \to \infty$. For this aim, we work with the extendible subfamily, via the Lévy-frailty construction, and we show how the probability distribution of S_d/d is closely related with the probability law of the exponential functional of Lévy subordinators.

References

1. Arbenz, P., Embrechts, P., Puccetti, G.: The AEP algorithm or the fast computation of the distribution of the sum of dependent random variables. Bernoulli-Bethesda **17**(2), 562–591 (2011)
2. Bennett, G.: Probability inequalities for the sum of independent random variables. J. Am. Stat. Assoc. **57**(297), 33–45 (1962)
3. Bertoin, J., Yor, M.: Exponential functionals of Lévy processes. Probab. Surv. **2**, 191–212 (2005)
4. Carmona, P., Petit, F., Yor, M.: Exponential Functionals of Lévy Processes. Lévy Processes. Birkhäuser, Boston (2001)
5. Cossette, H., Marceau, É.: The discrete-time risk model with correlated classes of business. Insur. Math. Econ. **26**(2), 133–149 (2000)
6. de Acosta, A.: Upper bounds for large deviations of dependent random vectors. Probab. Theory Relat. Fields **69**(4), 551–565 (1985)
7. Denuit, M., Genest, C., Marceau, É.: Stochastic bounds on sums of dependent risks. Insur. Math. Econ. **25**(1), 85–104 (1999)
8. Downton, F.: Bivariate exponential distributions in reliability theory. J. R. Stat. Soc. Ser. B (Methodological) **3**, 408–417 (1970)

9. Embrechts, P., Lindskog, F., McNeil, A.: Modelling dependence with copulas an applications to risk management. Handb. Heavy Tail. Distrib. Financ. **8**(329–384), 1 (2003)
10. Fang, K.T., Kotz, S., Ng, K.W.: Symmetric Multivariate and Related Distributions. Chapman and Hall, London (1990)
11. Finetti, B.D.: La prévision: ses lois logiques, ses sources subjectives. In: Annales de l'institut Henri Poincaré (1937)
12. Gaver, D.: Observing stochastic processes, and approximate transform inversion. Oper. Res. **14**(3), 444–459 (1966)
13. Giesecke, K.: A simple exponential model for dependent defaults. J. Fixed Income **13**(3), 74–83 (2003)
14. Gjessing, H., Paulsen, J.: Present value distributions with applications to ruin theory and stochastic equations. Stoch. Process. Appl. **71**(1), 123–144 (1997)
15. Kou, S., Wang, H.: First passage times of a jump diffusion process. Adv. Appl. Probab. **35**(2), 504–531 (2003)
16. Kuznetsov, A., Pardo, J.: Fluctuations of stable processes and exponential functionals of hypergeometric Lévy processes. Acta Appl. Math. **123**(1), 1–27 (2010)
17. Mai, J., Scherer, M.: Lévy-frailty copulas. J. Multivar. Anal. **100**(7), 1567–1585 (2009)
18. Mai, J., Scherer, M.: Reparameterizing Marshall-Olkin copulas with applications to sampling. J. Stat. Comput. Simul. **81**(1), 59–78 (2011)
19. Mai, J., Scherer, M.: Simulating Copulas: Stochastic Models, Sampling Algorithms, and Applications. Series in Quantitative Finance. Imperial College Press, London (2012). http://books.google.de/books?id=EzbRygAACAAJ
20. Marshall, A., Olkin, I.: A generalized bivariate exponential distribution. J. Appl. Probab. **4**(2), 291–302 (1967)
21. Marshall, A., Olkin, I.: A multivariate exponential distribution. J. Am. Stat. Assoc. **62**(317), 30–44 (1967)
22. Puccetti, G., Rüschendorf, L.: Sharp bounds for sums of dependent risks. J. Appl. Probab. **50**(1), 42–53 (2013)
23. Rivero, V.: Tail asymptotics for exponential functionals of Lévy processes: the convolution equivalent case. Probabilités et Statistiques, Annales de l'Institut Henri Poincaré (2012)
24. Stehfest, H.: Algorithm 368: numerical inversion of Laplace transforms. Commun. ACM **13**(1), 47–49 (1970)
25. Wüthrich, M.: Asymptotic value-at-risk estimates for sums of dependent random variables. Astin Bull. **33**(1), 75–92 (2003)

Chapter 4
General Marshall–Olkin Models, Dependence Orders, and Comparisons of Environmental Processes

Esther Frostig and Franco Pellerey

Abstract In many applicative fields, the behavior of a process $\mathbf{Z}$ is assumed to be subjected to an underlying process Θ that describes evolutions of environmental conditions. A common way to define the environmental process is by letting the marginal values of Θ subjected to specific environmental factors (constant along time) and factors describing the conditions of the environment at the specified time. In this paper we describe some recent results that can be used to compare two of such environmental discrete-time processes Θ and $\widetilde{\Theta}$ in dependence. A sample of applications of the effects of these comparison results on the corresponding processes $\mathbf{Z}$ and $\widetilde{\mathbf{Z}}$ in some different applicative contexts are provided.

4.1 Introduction

Assume that the behavior of a process $\mathbf{Z}$ is subjected to an underlying process Θ describing evolutions of environmental conditions. For example, the rate of growth in a population, or the wear accumulated by an item, or the total claim amount accumulated by an insurance company, can depend on random parameters describing the environment that evolves in time according to some process able to take into account both random factors due to the specific environment and random factors due to temporary conditions. In the discrete time case, a common way to define the environmental process $\Theta = \{\Theta_i, i \in \mathbb{N}^+\}$ by letting $\Theta_i = X_0 \odot X_i$, where $\odot$ is any binary operator, the variable X_0 describes the specific environmental factor, which is constant along the time, while the sequence $\{X_i, i = 1, 2, \ldots\}$, composed

We sincerely thank the referee for the careful reading of the paper.

E. Frostig
Department of Statistics, University of Haifa, Haifa, Israel
e-mail: frostig@stat.haifa.ac.il

F. Pellerey (✉)
Dipartimento di Scienze Matematiche, Politecnico di Torino, Turin, Italy
e-mail: franco.pellerey@polito.it

U. Cherubini et al. (eds.), *Marshall–Olkin Distributions - Advances in Theory and Applications*, Springer Proceedings in Mathematics & Statistics 141,
DOI 10.1007/978-3-319-19039-6_4

by independent and identically distributed variables, gives the additional description about the conditions of the environment at the fixed times i.

In this setup, it is sometimes of importance to figure out the influence of the common factor X_0 on the strength of positive dependence between the values assumed by the process Θ along time, which, in turn, reflects on the behavior of the process $\mathbf{Z}$ and on quantities related to this process like hitting times, or maximum reached values, etc.

One possible tool to make this analysis is by means of dependence orders, providing results able to evaluate the effects of the environmental factors on the intrinsic dependence in the evolutions of the environmental process. For this purpose, here we consider two environmental processes Θ and $\widetilde{\Theta}$ having the same marginal distributions and defined by letting

$$\Theta_i = X_0 \odot X_i \quad \text{and} \quad \widetilde{\Theta}_i = \widetilde{X}_0 \odot \widetilde{X}_i \quad \text{for all } i \in \mathbb{N},$$

for different environmental factors $X_0, X_i, \widetilde{X}_0$ and $\widetilde{X}_i$, with $i \in \mathbb{N}$, and then we describe conditions on the factors such that Θ and $\widetilde{\Theta}$ are comparable in positive dependence. Such dependence comparisons can be further applied to provide useful inequalities related to the corresponding processes $\mathbf{Z}$ and $\widetilde{\mathbf{Z}}$, as shown in the last section.

This is the plan of the paper. Section 4.2 is devoted to the descriptions of the dependence orders that will be used to compare the processes and to recall the definition of other useful stochastic orders. Also, a brief description of the general model used to define the environmental processes Θ and $\widetilde{\Theta}$ is given. The list of the existing results, useful to compare the two processes, is provided in Sect. 4.3, while Sect. 4.4 is devoted to applications of these results.

Some conventions and notations that are used throughout the paper are given below. The notation $=_{st}$ stands for equality in distribution. For any family of parameterized random variables $\{X_\theta \,|\theta \in \mathcal{T}\}$, such that $\mathcal{T} \subseteq \mathbb{R}$ is the support of a random variable Θ, then we denote by $X(\Theta)$ the mixture of the family $\{X_\theta \,|\theta \in \mathcal{T}\}$ with respect to Θ. For any random variable (or vector) X and an event A, $[X\,|A\,]$ denotes a random variable whose distribution is the conditional distribution of X given A. Also, throughout this paper we write "increasing" instead of "non-decreasing" and "decreasing" instead of "non-increasing".

4.2 Preliminaries

Useful definitions are recalled in this section. First, we recall the most well-known orders considered in the literature to compare the degree of positive dependence in components of random vectors.

4.2.1 Dependence Orders

Let $\prec$ denotes the coordinatewise ordering in $\mathbb{R}^n$. Given a function $\varphi : \mathbb{R}^n \to \mathbb{R}$, we recall that it is said to be *supermodular* if for any $\mathbf{x}, \mathbf{y} \in \mathbb{R}^n$ it satisfies

$$\varphi(\mathbf{x}) + \varphi(\mathbf{y}) \prec \varphi(\mathbf{x} \wedge \mathbf{y}) + \varphi(\mathbf{x} \vee \mathbf{y}), \tag{4.1}$$

where the operators $\wedge$ and $\vee$ denote, respectively, coordinatewise minimum and maximum.

Definition 4.1 Let $\mathbf{T} = (T_1, T_2, \ldots, T_n)$ and $\hat{\mathbf{T}} = (\hat{T}_1, \hat{T}_2, \ldots, \hat{T}_n)$ be two n-dimensional random vectors, then $\mathbf{T}$ is said to be smaller than $\hat{\mathbf{T}}$ in the *supermodular order* (denoted by $\mathbf{T} \prec_{\mathrm{sm}} \hat{\mathbf{T}}$) if

$$\mathbb{E}[\phi(\mathbf{T})] \leq \mathbb{E}[\phi(\hat{\mathbf{T}})],$$

for every supermodular real-valued function ϕ defined on $\mathbb{R}^n$ for which the expectations exist.

The supermodular order has been considered in several applied contexts (see [6, 13, 19, 20], or [1], among others). For a complete description of the supermodular order and its properties see [21].

A dependence order which is implied by the supermodular order (and equivalent to the supermodular order in the case $n = 2$) is the well-known positive orthant dependence order, whose definition is recalled next (see [21] for details).

Definition 4.2 Let $\mathbf{T} = (T_1, T_2, \ldots, T_n)$ and $\hat{\mathbf{T}} = (\hat{T}_1, \hat{T}_2, \ldots, \hat{T}_n)$ be two n-dimensional random vectors, then $\mathbf{T}$ is said to be smaller than $\hat{\mathbf{T}}$ in the *positive orthant dependence order* (denoted by $\mathbf{T} \prec_{\mathrm{POD}} \hat{\mathbf{T}}$) if

$$P[\mathbf{T} \leq \mathbf{x}] \leq P[\hat{\mathbf{T}} \leq \mathbf{x}] \quad \text{and} \quad P[\mathbf{T} > \mathbf{x}] \leq P[\hat{\mathbf{T}} > \mathbf{x}]$$

for every $\mathbf{x} \in \mathbb{R}^n$.

The supermodular order and the positive orthant dependence order can compare the dependence only in vectors having the same marginal distributions. Dealing with dependence in vectors having different marginal distributions, thus in different Fréchet classes, the comparison which is commonly considered is the *concordance* order, whose definition is based on the notion of copula, briefly recalled here.

Given the vector $\mathbf{T} = (T_1, T_2, \ldots, T_n)$, having joint distribution $F_{\mathbf{T}}$ and marginal distributions $F_1, \ldots, F_n$, the function $C_{\mathbf{T}} : [0, 1]^n \to [0, 1]$ satisfying

$$F_{\mathbf{T}}(x_1, \ldots, x_n) = C_{\mathbf{T}}(F_1(x_1), \ldots, F_n(x_n)), \quad \forall (x_1, \ldots, x_n) \in \mathbb{R}^n$$

is said to be the *copula* of $\mathbf{T}$. If the marginal distributions F_i, for $i = 1, \dots, n$, are continuous, then the copula $C_{\mathbf{T}}$ is unique and it is defined as

$$C_{\mathbf{T}}(u_1, \dots, u_n) = F_{\mathbf{T}}(F_1^{-1}(u_1), \dots, F_n^{-1}(u_n)) = P[F_1(T_1) \leq u_1, \dots, F_n(T_n) \leq u_n],$$

where F_i^{-1} denotes the right continuous inverse of F_i, i.e., $F_i^{-1}(u) = \sup\{x : F_i(x) \leq u\}$, $u \in [0, 1]$.

Copulas entirely describe the dependence between the components of a random vector; for example, concordance indexes like the Spearman's ρ or Kendall's τ of a vector $\mathbf{T}$ can be defined by means of its copula (see [15] for a monograph on copulas and their properties). In particular, the following dependence order is defined by comparison of copulas.

Definition 4.3 Let $\mathbf{T} = (T_1, T_2, \dots, T_n)$ and $\hat{\mathbf{T}} = (\hat{T}_1, \hat{T}_2, \dots, \hat{T}_n)$ be two n-dimensional random vectors having copulas $C_{\mathbf{T}}$ and $C_{\hat{\mathbf{T}}}$, respectively. Then $\mathbf{T}$ is said to be smaller than $\hat{\mathbf{T}}$ in the *concordance order* (denoted by $\mathbf{T} \prec_{\mathrm{c}} \hat{\mathbf{T}}$) if

$$C_{\mathbf{T}}(u_1, \dots, u_n) \leq C_{\hat{\mathbf{T}}}(u_1, \dots, u_n) \quad \forall (u_1, \dots, u_n) \in [0, 1]^n.$$

It is useful to observe that, in order to be comparable in concordance, the vectors $\mathbf{T}$ and $\hat{\mathbf{T}}$ do not need to have the same marginal distributions. However, when they have the same continuous marginal distributions $F_1, \dots, F_n$, one can consider the random vectors $\mathbf{U}$ and $\mathbf{V}$ defined as

$$\mathbf{U} = (F_1^{-1}(T_1), \dots, F_n^{-1}(T_n)) \quad \text{and} \quad \mathbf{V} = (F_1^{-1}(\hat{T}_1), \dots, F_n^{-1}(\hat{T}_n)),$$

and observe that their distribution functions are the copulas $C_{\mathbf{T}}$ and $C_{\hat{\mathbf{T}}}$ of $\mathbf{T}$ and $\hat{\mathbf{T}}$, respectively. Applying Theorem 9.A.9(a) in [21] immediately follows that $\mathbf{X} \prec_{\mathrm{sm}} \mathbf{Y}$ iff $\mathbf{U} \prec_{\mathrm{sm}} \mathbf{V}$. Observing that the latter implies $\mathbf{U} \prec_{\mathrm{POD}} \mathbf{V}$, which in turn is equivalent to $C_{\mathbf{T}}(u_1, \dots, u_n) \leq C_{\hat{\mathbf{T}}}(u_1, \dots, u_n)$, for all $(u_1, \dots, u_n) \in [0, 1]^n$, the following assertion holds.

Proposition 4.1 *Assume that* $\mathbf{T} = (T_1, \dots, T_n)$ *and* $\hat{\mathbf{T}} = (\hat{T}_1, \dots, \hat{T}_n)$ *are two random vectors having the same continuous marginal distributions. If* $\mathbf{T} \prec_{sm} \hat{\mathbf{T}}$, *then also* $\mathbf{T} \prec_c \hat{\mathbf{T}}$.

We also recall that comparisons between random pairs based on their intrinsic degree of dependence can be defined by considering concordance indexes, in particular, by using the Kendall's τ concordance coefficient, whose definition is recalled here. For it, let (T_1, T_2) be a random vector and consider also a copy (T_1', T_2') that is independent of (T_1, T_2); that is, $(T_1', T_2') =_{\mathrm{st}} (T_1, T_2)$. Then $\tau_{(T_1,T_2)}$ is defined as follows:

$$\tau_{(T_1,T_2)} = 2P\{(T_1 - T_1')(T_2 - T_2') \geq 0\} - 1.$$

It is a well-known fact that

$$(T_1, T_2) \prec_c (\hat{T}_1, \hat{T}_2) \Rightarrow \tau_{(T_1,T_2)} \le \tau_{(\hat{T}_1,\hat{T}_2)}.$$

More details about this coefficient can be found, for instance, in [2] or [15].

4.2.2 Some Univariate Stochastic Orders

For the applications, it will be useful to consider some well-known univariate stochastic orders, whose definitions are recalled here. For it, let X_1 and X_2 be two random variables, having cumulative distribution functions F_1 and F_2, respectively, and denote with F_1^{-1} and F_2^{-1} the respective right continuous inverses.

Definition 4.4 X_1 is said to be smaller than X_2 in the

(a) *usual stochastic order* (denoted by $X_1 \le_{st} X_2$) if $E[\phi(X_1)] \le E[\phi(X_2)]$ for every increasing function ϕ such that the expectations exist;
(b) *increasing convex order* (denoted by $X_1 \le_{icx} X_2$) if $E[\phi(X_1)] \le E[\phi(X_2)]$ for every increasing and convex function ϕ such that the expectations exist;
(c) *dispersive order* (denoted by $X_1 \le_d X_2$) if $F_1^{-1}(\beta) - F_1^{-1}(\alpha) \le F_2^{-1}(\beta) - F_2^{-1}(\alpha)$, for all $0 < \alpha < \beta < 1$.

See [14] or [21] for thorough studies of these orders, and for the well-known usual stochastic order in particular. Here we just point out that the increasing convex order, also known as *stop-loss order*, is a stochastic comparison often considered in actuarial sciences and reliability. The main reason of its interest in actuarial sciences is that two risks are ordered in this sense if and only if the corresponding stop-loss transforms are ordered for every threshold t, i.e., $X_1 \le_{icx} X_2$ if and only if $E[(X_1 - t)^+] \le E[(X_2 - t)^+]$, for all $t \in \mathbb{R}$, where $x^+ = \max\{0, x\}$. For a comprehensive discussion on properties and applications of the increasing convex ordering we refer to [21] or [13].

For what it concerns the dispersive order, it should be pointed out that it is a well-known comparison of random variables based on their variability (see, again, [21])

4.2.3 The General Model for the Environmental Processes

The model presented here to define the environmental processes is a natural generalization, in higher dimensions, of the one defined and studied in the bivariate setting in [12]. We introduce the model considering vectors with a finite number of components; the case of discrete processes is a further generalization for higher numbers of components.

Let $\odot$ denotes any binary increasing, commutative and associative operator between real numbers, i.e., such that $x \odot y = y \odot x$ and $(x \odot y) \odot z = x \odot (y \odot z)$, for all $x, y, z \in \mathbb{R}$, and denote with $\bigodot$ the repeated application of this operator: $\bigodot_{i\in\{1,2,\ldots,n\}} x_i = x_1 \odot x_2 \odot \cdots \odot x_n$. For example, the operator $\odot$ can be minimum $\wedge$ or the maximum $\vee$, and in this case $\bigodot_{i\in\{1,2,\ldots,n\}} x_i = \bigwedge_{i\in\{1,2,\ldots,n\}} x_i$ or $\bigodot_{i\in\{1,2,\ldots,n\}} x_i = \bigvee_{i\in\{1,2,\ldots,n\}} x_i$, respectively, or it can be the sum or the product (restricted to non-negative real numbers, to preserve increasing property), so that in these cases $\bigodot_{i\in\{1,2,\ldots,n\}} x_i$ stands for $\sum_{i\in\{1,2,\ldots,n\}} x_i$ or $\prod_{i\in\{1,2,\ldots,n\}} x_i$.

Let $I = \{1, \ldots, n\}$ and $\mathbf{S} = \{S_j, j \in J \subseteq \mathbb{N}\}$ be a collection of subsets of I. Also, let $\{X_j, j \in J\}$ be a set of independent random variables describing possible factors influencing multivariate risks or lifetimes. Define, for $i = 1, \ldots, n$, the set $\Lambda_i = \{S_j : i \in S_j\}$ and let $T_i = \bigodot_{\{j:S_j\in\Lambda_i\}} X_j$ be the i-th component in the vector $\mathbf{T} = (T_1, \ldots, T_n)$. For example, given $I = \{1, 2\}$ and $\mathbf{S} = \{S_1, S_2, S_3\} = \{\{1\}, \{2\}, \{1, 2\}\}$, one can consider the independent variables X_1, X_2, X_3 to define the vector of lifetimes $\mathbf{T} = (T_1, T_2) = (X_1 \odot X_3, X_2 \odot X_3)$, modeling the fact that the first lifetime is influenced by factors X_1 and X_3, while factors X_2 and X_3 act on the second lifetime. In this example X_1 and X_2 describe individual factors (that influence the first and the second components, respectively), while X_3 is a random factor acting on both the components of $\mathbf{T}$.

This model has been considered in reliability theory to describe the vector of lifetimes of a set of components subjected to common and individual shocks. In fact, one can think at the X_j as the waiting time to type j shock, that causes the failure of components indexed by S_j. Defining $\Lambda_i = \{S_j : i \in S_j\}$, for $i = 1, \ldots, n$, and letting $\odot$ to be the minimum, so that $T_i = \min_{j:S_j\in\Lambda_i}\{X_j\}$ is the time to failure of component i, the joint distribution of $\mathbf{T} = (T_1, \ldots, T_n)$ is the *Generalized Marshall–Olkin (GMO)* distribution studied in [9] (see also the references therein). In the particular case where the X_j are exponentially distributed, it reduces to the well-known multivariate exponential distribution defined by [11]. Similar models have been considered also in [22] and [23], where applications in queueing systems are provided.

Also, similar models have been considered in risk theory or in multiple default problems to describe sets of dependent risks. Indeed, let again $\mathbf{S} = \{S_j, j \in J \subseteq \mathbb{N}\}$ be a collection of subsets of $I = \{1, \ldots, n\}$, and let $\{X_j, j \in J\}$ be a set of independent random variables. Assume that every X_j additively acts on all the components of index $i \in S_j$. Define, for all $i = 1, \ldots, n$, the set $\Lambda_i = \{S_j : i \in S_j\}$. Then one can consider the vector $\mathbf{T} = (T_1, \ldots, T_n)$ of dependent risks, where $T_i = \sum_{j:S_j\in\Lambda_i} X_j$, for all $i = 1, \ldots, n$, thus replacing the operator $\odot$ with the sum.

4.3 Comparison Results

Let us consider a random vector $\mathbf{T} = (T_1, T_2) = (X_1 \odot X_3, X_2 \odot X_3)$ defined as described in Sect. 4.2.3, and consider a new random vector $\widetilde{\mathbf{T}} = (\widetilde{T}_1, \widetilde{T}_2) = (\widetilde{X}_1 \odot$

$\widetilde{X}_3, \widetilde{X}_2 \odot \widetilde{X}_3)$ defined by means of the same set $\mathbf{S} = \{S_1, S_2, S_3\} = \{\{1\}, \{2\}, \{1, 2\}\}$, but considering different common and individual random factors $\widetilde{X}_j$, $j = 1, 2, 3$.

For particular choices of the operator $\odot$ the following results have been proved in recent literature.

Proposition 4.2 *Let* $X_1 \leq_{st} \widetilde{X}_1$, $X_2 \leq_{st} \widetilde{X}_2$ *and* $X_3 \geq_{st} \widetilde{X}_3$. *Then*

$$(X_1 \wedge X_3, X_2 \wedge X_3) \prec_c (\widetilde{X}_1 \wedge \widetilde{X}_3, \widetilde{X}_2 \wedge \widetilde{X}_3).$$

This statement, whose proof may be found in [9], essentially says that the positive dependence among the vector's components increases, in concordance, as the common factor X_3 stochastically decreases. Further generalizations of this result, in higher dimensions or considering the maximum instead of the minimum, can be found in [10] and [3].

When the operator $\odot$ is the sum, then the positive dependence between the components of the vector (evaluated in terms of the Kendall's τ concordance coefficient) increases as the variability of the common factor increases. A result in this direction is the following, whose proof may be found in [17].

Proposition 4.3 *Let* $X_1, X_2, \widetilde{X}_1$ *and* $\widetilde{X}_2$ *be independent and identically distributed, and let* $X_3 \leq_d \widetilde{X}_3$. *Then*

$$\tau_{(X_1+X_3, X_2+X_3)} \leq \tau_{(\widetilde{X}_1+\widetilde{X}_3, \widetilde{X}_2+\widetilde{X}_3)}.$$

Other similar results, dealing with the additive model, are provided in [17].

In order to provide comparisons of positive dependence in a more general framework, new results, dealing with the supermodular order, have been proved in [7]. Here we recall the main result contained in that paper.

For it, consider a vector $\mathbf{T} = (T_1, \ldots, T_n)$ defined as described in Sect. 4.2.3. Then, consider a set $\mathbf{I} = \{I_1, \ldots, I_k\} \subseteq \mathbf{S}$ be such that the I_r, $r = 1, \ldots, k$, are disjoint subsets of $I = \{1, \ldots, n\}$ and also such that $\bigcup_{r=1}^k I_r \in \mathbf{S}$, and let $\mathcal{E}$ and $\mathcal{E}_r$, $r = 1, \ldots, k$ be independent and identically distributed random variables, independent of the X_j and $\widetilde{X}_j$. Let:

$$\begin{array}{lll} \widetilde{X}_j & =_{st} X_j & \text{if } S_j \notin \{I_1, \ldots, I_k, \cup_{r=1}^k I_r\}, \\ X_j & =_{st} \widetilde{X}_j \odot \mathcal{E} & \text{if } S_j = \cup_{r=1}^k I_r, \\ \widetilde{X}_j & =_{st} X_j \odot \mathcal{E}_r & \text{if } S_j \in \{I_1, \ldots, I_k\}. \end{array}$$

Now consider the random vector $\widetilde{\mathbf{T}} = (\widetilde{T}_1, \ldots, \widetilde{T}_n)$, where the $\widetilde{T}_i$ are defined as $\widetilde{T}_i = \bigodot_{\{j: S_j \in \Lambda_i\}} \widetilde{X}_j$. It is easy to observe that, under relations above, because of associativity of $\odot$, the vectors $\mathbf{T} = (T_1, \ldots, T_n)$ and $\widetilde{\mathbf{T}} = (\widetilde{T}_1, \ldots, \widetilde{T}_n)$ have the same marginal distributions.

For example, given $\mathbf{T} = (T_1, T_2) = (X_1 \odot X_3, X_2 \odot X_3)$, a new vector having the same marginal distributions of $\mathbf{T}$ can be defined considering three new independent and identically distributed random variables $\mathcal{E}$, $\mathcal{E}_1$, and $\mathcal{E}_2$, observing that $\mathbf{T}$ is defined by letting $\mathbf{S} = \{S_1, S_2, S_3\} = \{\{1\}, \{2\}, \{1, 2\}\}$, and considering the independent variables $\widetilde{X}_j$ such that $\widetilde{X}_1 = X_1 \odot \mathcal{E}_1$, $\widetilde{X}_2 = X_2 \odot \mathcal{E}_2$ and $X_3 = \widetilde{X}_3 \odot \mathcal{E}$. The corresponding vector is $\widetilde{\mathbf{T}} = (\widetilde{T}_1, \widetilde{T}_2) = (\widetilde{X}_1 \odot \widetilde{X}_3, \widetilde{X}_2 \odot \widetilde{X}_3)$, and the two vectors $\mathbf{T}$ and $\widetilde{\mathbf{T}}$ have the same marginal distributions, being, e.g.,

$$T_1 = X_1 \odot X_3 = X_1 \odot \mathcal{E} \odot \widetilde{X}_3 =_{st} X_1 \odot \mathcal{E}_1 \odot \widetilde{X}_3 = \widetilde{X}_1 \odot \widetilde{X}_3 = \widetilde{T}_1.$$

The following statement provides a comparison of the degree of dependence in the two vectors.

Proposition 4.4 *Let the vectors* $\mathbf{T}$ *and* $\widetilde{\mathbf{T}}$ *be defined as above. Then* $\widetilde{\mathbf{T}} \prec_{sm} \mathbf{T}$.

We refer the reader to [7] for the proof and for other similar results.

Note that the previous statements can be restated in terms of concordance order.

Corollary 4.1 *Let the vectors* $\widetilde{\mathbf{T}}$ *and* $\mathbf{T}$ *be defined as Proposition 4.4. Assume they have continuous marginal distributions. Then* $\widetilde{\mathbf{T}} \prec_c \mathbf{T}$.

4.4 Supermodular Comparison of Environmental Processes

Consider two different environmental processes Θ and $\widetilde{\Theta}$ having the same marginal distributions, defined as in Sect. 4.1 by letting

$$\Theta_i = X_0 \odot X_i \quad \text{and} \quad \widetilde{\Theta}_i = \widetilde{X}_0 \odot \widetilde{X}_i \quad \text{for all } i \in \mathbb{N}. \tag{4.2}$$

Let $\mathcal{E}$ and $\mathcal{E}_i$, $i \in \mathbb{N}$, be a set of independent and identically distributed random variables. If

$$X_0 =_{st} \widetilde{X}_0 \odot \mathcal{E} \text{ and } \widetilde{X}_i =_{st} X_i \odot \mathcal{E}_i \text{ for all } i = 1, 2, \ldots, \tag{4.3}$$

i.e., if the marginal distributions of the processes are the same, then in the first case the specific environmental random factor has a higher impact on the dependence than in the second case. In fact, through Proposition 4.4 one gets

$$(\widetilde{\Theta}_1, \ldots, \widetilde{\Theta}_n) \prec_{\text{sm}} (\Theta_1, \ldots, \Theta_n) \; \forall n \in \mathbb{N}^+.$$

Such a dependence comparison can be further applied to provide useful inequalities, as shown in the following examples.

4.4.1 An Application in Population Dynamics

Consider a branching process that describes the growth of a population, and assume that it depends on the environment as follows. Let $\Theta = \{\theta_0, \theta_1, \ldots, \}$ be a sequence of values in a set $\mathcal{T}$ describing the evolutions of the environment, and define, recursively, the stochastic process $\mathbf{Z}(\Theta) = \{Z_n(\theta_0, \dots, \theta_n), n \in \mathbb{N}\}$ by

$$Z_0(\theta_0) = Y_{1,0}(\theta_0)$$

and

$$Z_n(\theta_0, \dots, \theta_n) = \sum_{j=1}^{Z_{n-1}(\theta_0,\dots,\theta_{n-1})} Y_{j,n}(\theta_n), \quad n \geq 1.$$

In order to deal with random evolutions of the environment, consider a sequence $\Theta = (\Theta_0, \Theta_1, \dots)$ of random variables taking on values in $\mathcal{T}$ and consider the stochastic process $\mathbf{Z}(\Theta) = \{Z_n(\Theta_0, \dots, \Theta_n), n \in \mathbb{N}\}$ defined by

$$Z_0(\Theta_0) = Y_{1,0}(\Theta_0)$$

and

$$Z_n(\Theta_0, \dots, \Theta_n) = \sum_{j=1}^{Z_{n-1}(\Theta_0,\dots,\Theta_{n-1})} Y_{j,n}(\Theta_n), \quad n \geq 1,$$

where, for every $j, k \in \mathbb{N}$, $Y_{j,k}(\Theta_k)$ is a nonnegative random variable such that

$$[Y_{j,k}(\Theta_k)|\Theta_k = \theta] =_{st} Y_{j,k}(\theta).$$

Assume the $Y_{j,k}(\theta_k)$ to be sequences of independent and identically distributed variables, for every fixed θ_k, and assume that they are stochastically increasing in θ_k.

Now consider two different environmental processes Θ and $\widetilde{\Theta}$ as defined previously, letting for example the operator $\odot$ to be the sum (to describe the fact that adverse conditions act additively). If $X_0 =_{st} \widetilde{X}_0 + \mathcal{E}$ and $\widetilde{X}_i =_{st} X_i + \mathcal{E}$ for all $i = 1, 2, \dots$, i.e., if in the first case, the constant environmental random factor X_0 is stochastically greater than the constant environmental random factor $\widetilde{X}_0$ of the second case, then $\widetilde{\Theta} \prec_{\text{sm}} \Theta$. Observing that all the assumptions of Proposition 4.2 in [4] are satisfied, and recalling that the supermodular order implies the increasing directionally convex order, by Corollary 4.2 in [4] immediately follows

$$Z_n(\widetilde{\Theta}_1, \dots, \widetilde{\Theta}_n) \leq_{icx} Z_n(\Theta_1, \dots, \Theta_n) \quad \forall\, n \in \mathbb{N}^+$$

(see [4] for details on the increasing directionally convex order). This means that the numbers of individuals in the two populations can be ordered according to the increasing convex order at any fixed time n, i.e., the number of individuals in the population increases in the *icx* order as the constant random factor describing the environment stochastically decreases. It should be pointed out that from this comparison it also follows that the probabilities of extinction (for fixed n) are comparable, as shown in Corollary 8.6.8 in [18].

4.4.2 An Application in Comparisons of Collective Risks

Consider a portfolio of n risks over a single period of time and assume that during that period each policyholder i can have a nonnegative claim Y_i with probability $\theta_i \in [0, 1] \subseteq \mathbb{R}$. Then, the total claim amount $S(\theta_1, \ldots, \theta_n)$ during that time can be represented as

$$S(\theta_1, \ldots, \theta_n) = \sum_{i=1}^{n} I_i(\theta_i) Y_i,$$

where $I_i(\theta_i)$ denotes a Bernoulli random variable with parameter θ_i.

In order to remove the assumption of independence among the Bernoulli random variables, one can replace the vector of real parameters $(\theta_1, \ldots, \theta_n)$ by a random vector $\Theta = (\Theta_1, \ldots, \Theta_n)$, with values in $[0, 1]^n \subseteq \mathbb{R}^n$, describing both the random environment for occurrences of claims and the dependence among them (see, e.g., [5]).

Assume that the risks Y_i are independent and identically distributed, and assume that every component Θ_i of Θ is the maximum among two random parameters X_0 and X_i assuming values in [0, 1], describing, respectively, a cause of risk common to all policyholders and the individual propensity of risk. Thus Θ can be defined with the model described in Sect. 4.2, letting $\odot$ be the maximum. Consider now a different environment $\widetilde{\Theta}$ similarly defined, with common and individual factors $\widetilde{X}_0$ and $\widetilde{X}_i$, and assume that there exist a set of independent and identically distributed random variables $\mathcal{E}$ and $\mathcal{E}_i$, $i \in \mathbb{N}$ such that

$$X_0 =_{st} \widetilde{X}_0 \vee \mathcal{E} \;\text{ and }\; \widetilde{X}_i =_{st} X_i \vee \mathcal{E}_i \;\text{ for all }\; i = 1, 2, \ldots,$$

This assumption can model, for example, the fact that the common cause of risk is stochastically greater in the environment Θ than in $\widetilde{\Theta}$.

By applying Proposition 4.4 one immediately gets $\widetilde{\Theta} \prec_{\text{sm}} \Theta$. Now, from Corollary 4.1 in [4], observing that its assumptions are satisfied whenever the risks Y_i are independent and identically distributed, and again recalling that the supermodular order implies the increasing and directionally convex order, it follows $S(\widetilde{\Theta}) \leq_{icx} S(\Theta)$. Comparisons of stop losses, or premiums, for the collective risks under the two environments follow from the last stochastic comparison. For example,

$$E[(S(\widetilde{\Theta}) - \alpha)^+] \leq E[(S(\Theta) - \alpha)^+] \;\text{ for all } \alpha \in \mathbb{R}.$$

4.4.3 An Application in Ruin Theory

Consider the reserve of an insurance company along time, $R(t) = u + \beta t - S(t)$, with $t \geq 0$, where $S(t)$ is a claim process, β is the rate company get paid per time unit, and u is the original capital the company owns. In general, it is assumed that $\beta > E[Z_i]$, where Z_i represents the amount of claims in unit time.

Consider the corresponding probability of ruin $P[\max(S(t) - \beta t) \geq u]$. It is a well-known fact that to evaluate the probability of ruin, the Lundberg approximation

$$P[\max_{s \in (0,t)} (S(s) - \beta s) \geq u] \overset{t \to \infty}{\sim} \mathrm{e}^{-\gamma u},$$

where the *Lundberg coefficient* γ satisfies the equation $\ln(E[\mathrm{e}^{\gamma X_i}]) = \beta$.

Let us now consider a discrete time version of this model, letting $S_n = \Sigma_{i=1}^n Z_i$, where the Z_i are non-necessarily independent and identically distributed. It has been shown in [16], that in this case

$$P[\max_{n \in \mathbb{N}}(S_n - \beta n) \geq u] \sim \mathrm{e}^{-\gamma u},$$

where γ satisfies

$$\lim_{n \to \infty} \frac{E[\mathrm{e}^{\gamma S_n - \beta n}]}{n} = 0 \text{ or, equivalently, } \lim_{n \to \infty} \frac{E[\mathrm{e}^{\gamma S_n}]}{n} = \beta,$$

when the limit exists.

Consider now an environmental process $\Theta = \{\Theta_1, \Theta_2, \ldots\}$ and a claim process $\mathbf{Z} = \{Z_1(\Theta_1), Z_2(\Theta_2), \ldots\}$ which depends on the environmental process. Assume the claims $Z_i(\theta_i)$ to be independent for fixed values of the parameters θ_i, and stochastically increasing in θ_i. Consider now a new environmental process $\widetilde{\Theta} = \{\widetilde{\Theta}_1, \widetilde{\Theta}_2, \ldots\}$, and assume that both the processes depend on a common factor constant along the time, say X_0 and $\widetilde{X}_0$, respectively, and factors describing instantaneous conditions, say X_i and $\widetilde{X}_i$. Assume there exists $\mathcal{E}$ and $\mathcal{E}_i$, $i \in \mathbb{N}$ such that Θ and $\widetilde{\Theta}$ satisfy Eqs. (4.2) and (4.3) for a suitable operator $\odot$ (like, for example, the sum, to model the fact that factors act additively. Using the results described in the previous section we get $(\widetilde{\Theta}_1, \widetilde{\Theta}_2, \ldots, \widetilde{\Theta}_n) \prec_{\text{sm}} (\Theta_1, \Theta_2, \ldots, \Theta_n)$, and therefore also, by the closure under mixture of the supermodular order,

$$(Z_1(\widetilde{\Theta}_1), \ldots, Z_n(\widetilde{\Theta}_n)) \prec_{\text{sm}} (Z_1(\Theta_1), \ldots, Z_n(\Theta_n)),$$

and, as a consequence,

$$\widetilde{S}_n = \Sigma_{i=1}^n Z_i(\widetilde{\Theta}_i) \leq_{icx} S_n = \Sigma_{i=1}^n Z_i(\Theta_i)$$

(see [21], for details).

Let $\widetilde{\gamma}$ be the Lundberg coefficient for the second process, i.e., let $\lim_{n\to\infty} \dfrac{E[\mathrm{e}^{\widetilde{\gamma}\widetilde{S}_n}]}{n} = \beta$.

Since $\widetilde{S}_n \leq_{icx} S_n$, then, by definition of the increasing convex order, $E[\phi(\widetilde{S}_n)] \leq E[\phi(S_n)]$ for every increasing and convex function ϕ (for which the expectations exist). Thus, in particular,

$$\frac{E[\mathrm{e}^{\widetilde{\gamma}\widetilde{S}_n}]}{n} \leq \frac{E[\mathrm{e}^{\widetilde{\gamma}S_n}]}{n} \quad \forall n \in \mathbb{N}^+$$

and

$$\beta = \lim_{n\to\infty} \frac{E[\mathrm{e}^{\widetilde{\gamma}\widetilde{S}_n}]}{n} \leq \lim_{n\to\infty} \frac{E[\mathrm{e}^{\widetilde{\gamma}S_n}]}{n}.$$

Since γ should satisfy $\lim_{n\to\infty} \dfrac{E[\mathrm{e}^{\gamma S_n}]}{n} = \beta$, it follows that $\gamma \leq \widetilde{\gamma}$.

Thus,

$$P[\max_{n\in\mathbb{N}}(S_n - \beta n) \geq u] \sim \mathrm{e}^{-\gamma u} \geq \mathrm{e}^{-\widetilde{\gamma} u} \sim P[\max_{n\in\mathbb{N}}(\widetilde{S}_n - \beta n) \geq u] \;\; \forall u \in \mathbb{R}^+,$$

namely, the second case has a higher ruin probability than that of the first case, i.e., the probability of ruin increases as the systematic random factor describing the environment stochastically decreases (whenever the premium rate β is fixed).

4.4.4 An Application in Reliability

Consider a set $\{\mathbf{N}_j = \{N_j(t), t \geq 0\}, j \in J\}$ of independent Poisson processes. Then define a multivariate Poisson process $\mathbf{M} = (\mathbf{M}_1, \ldots, \mathbf{M}_n)$ like in the additive model, i.e., assume that every $\mathbf{M}_i$ is the superposition of some of the processes $\mathbf{N}_j$:

$$M_i(t) = \sum_{j:S_j \in \Lambda_i} N_j(t), \; t \geq 0,$$

for an appropriate choice of the set $\mathbf{S} = \{S_j, j \in J \subseteq \mathbb{N}\}$.

This kind of processes are commonly used to count the number of customers arriving in different lines of service, or the number of claims due to different causes, or the number of shocks occurring to components, etc.

Consider now a new multivariate Poisson process $\widetilde{\mathbf{M}} = (\widetilde{\mathbf{M}}_1, \ldots, \widetilde{\mathbf{M}}_n)$ defined as above by means of the same set $\mathbf{S}$ and the new independent Poisson processes $\{\widetilde{\mathbf{N}}_j = \{\widetilde{N}_j(t), t \geq 0\}, j \in J\}$. Let λ_j and $\widetilde{\lambda}_j$ denote the intensity rates of the processes $\mathbf{N}_j$ and $\widetilde{\mathbf{N}}_j$, respectively.

Let $\mathbf{I} = \{I_1, \ldots, I_k\} \subseteq \mathbf{S}$, be such that the $I_r, r = 1, \ldots, k$, are disjoint sets in $\mathbf{I}$ and also such that $\bigcup_{r=1}^k I_r \in \mathbf{S}$, and consider any fixed $\delta \in \mathbb{R}^+$ such that $\delta \leq \lambda_j, \ \forall j \in J$. Assume that

$$\widetilde{\lambda}_j = \begin{cases} \lambda_j & \text{if } S_j \notin \{I_1, \ldots, I_k, \cup_{r=1}^k I_r\}, \\ \lambda_j - \delta & \text{if } S_j = \cup_{r=1}^k I_r, \\ \lambda_j + \delta & \text{if } S_j \in \{I_1, \ldots, I_k\}. \end{cases}$$

Observing that $\widetilde{\mathbf{M}}$ has the same marginal distributions of $\mathbf{M}$, from Proposition 4.4 one immediately gets that, for every fixed time t,

$$(\widetilde{M}_1(t), \ldots, \widetilde{M}_n(t)) \leq_{sm} (M_1(t), \ldots, M_n(t)). \tag{4.4}$$

Consider now a series system, where each component (assume in position i) dies whenever it collects an amount $k + 1$ of shocks, arriving according to Poisson processes as above. Thus, component i is alive at time t if $M_i(t) \leq k$, and the whole system is broken at time t if $M(t) = \min\{M_i(t)\} > k$. Let P_t denotes the probability that the system is broken at time t, i.e., let $P_t = P[M(t) > k]$, and similarly define $\widetilde{P}_t$

Recall now that the minimum is an increasing supermodular function, and that the composition $\phi \circ \min$ is supermodular if the function ϕ is increasing (see, e.g., [8]). Thus, for any increasing function ϕ by (4.4) we have

$$\begin{aligned} E[\phi(\widetilde{M}(t))] &= E[\phi(\min\{\widetilde{M}_i(t), i = 1, \ldots n\})] \\ &\leq E[\phi(\min\{M_i(t), i = 1, \ldots n\})] = E[\phi(M(t))], \end{aligned}$$

i.e., $\widetilde{M}(t) \leq_{st} M(t)$, for all t. Also, $\widetilde{P}_t \leq P_t$, for all $t \geq 0$, i.e., the lifetime of the first system is stochastically greater than those of the second system.

References

1. Denuit, M., Frostig, E., Levikson, B.: Supermodular comparison of time-to-ruin random vectors. Methodol. Comput. Appl. Probab. **9**, 41–54 (2007)
2. Mari, D.D., Kotz, S.: Correlation and Dependence. Imperial College Press, London (2001)
3. Fang, R., Li, X.: A note on bivariate dual generalized Marshall-Olkin distributions with applications. Probab. Eng. Inf. Sci. **27**, 367–374 (2013)
4. Fernández-Ponce, J.M., Ortega, E., Pellerey, F.: Convex comparisons for random sums in random environments and application. Probab. Eng. Inf. Sci. **22**, 389–413 (2008)
5. Frostig, E.: Comparison of portfolios which depend on multivariate Bernoulli random variables with fixed marginals. Insur.: Math. Econ. **29**, 319–331 (2001)
6. Frostig, E.: Ordering ruin probabilities for dependent claim streams. Insur.: Math. Econ. **32**, 93–114 (2003)
7. Frostig, E., Pellerey, F.: A general framework for the supermodular comparisons of models incorporating individual and common factors. Technical Report, Department of Statistics, University of Haifa, Israel (2013)

8. Di Crescenzo, A., Frostig, E., Pellerey, F.: Stochastic comparisons of symmetric supermodular functions of heterogeneous random vectors. J. Appl. Probab. **50**, 464–474 (2013)
9. Li, X., Pellerey, F.: Generalized Marshall-Olkin distributions and related bivariate aging properties. J. Multivar. Anal. **102**, 1399–1409 (2011)
10. Lin, J., Li, X.: Multivariate generalized Marshall-Olkin distributions and copulas. Methodol. Comput. Appl. Probab. **16**, 53–78 (2012)
11. Marshall, A.W., Olkin, I.: A multivariate exponential distribution. J. Am. Stat. Assoc. **62**, 30–41 (1967)
12. Muliere, P., Scarsini, M.: Characterization of a Marshall-Olkin type class of distributions. Ann. Inst. Stat. Math. **39**, 429–441 (1987)
13. Müller, A.: Stop-loss order for portfolios of dependent risks. Insur.: Math. Econ. **21**, 219–223 (1997)
14. Müller, A., Stoyan, D.: Comparison Methods for Stochastic Models and Risks. Wiley, Chichester (2002)
15. Nelsen, R.B.: An Introduction to Copulas, 2nd edn. Springer, Berlin (2006)
16. Nyrhinen, H.: Rough descriptions of ruin for a general class of surplus processes. Adv. Appl. Probab. **30**, 1008–1026 (1998)
17. Pellerey, F., Shaked, M., Sekeh, S.Y.: Comparisons of concordance in additive models. Stat. Probab. Lett. **582**, 2059–2067 (2012)
18. Ross, S.M.: Stochastic Processes. Wiley, New York (1983)
19. Rüschendorf, L.: Comparison of multivariate risks and positive dependence. Adv. Appl. Probab. **41**, 391–406 (2004)
20. Shaked, M., Shanthikumar, J.G.: Supermodular stochastic orders and positive dependence of random vectors. J. Multivar. Anal. **61**, 86–101 (1997)
21. Shaked, M., Shanthikumar, J.G.: Stochastic Orders. Springer, New York (2007)
22. Xu, S.H.: Structural analysis of a queueing system with multi-classes of correlated arrivals and blocking. Oper. Res. **47**, 263–276 (1999)
23. Xu, S.H., Li, H.: Majorization of weighted trees: a new tool to study correlated stochastic systems. Math. Oper. Res. **25**, 298–323 (2000)

Chapter 5
Marshall–Olkin Machinery and Power Mixing: The Mixed Generalized Marshall–Olkin Distribution

Sabrina Mulinacci

Abstract In this paper, we consider the Marshall–Olkin technique of modeling the multivariate random lifetimes of the components of a system, as the first arrival times of some shock affecting part or the whole system and we analyze the possibility to add more dependence among the shocks and, as a consequence, among the lifetimes, through the power-mixing technique. This approach is applied to obtain extensions of the generalized Marshall–Olkin distributions.

5.1 Introduction

Following the Marshall and Olkin [18] seminal paper, the multivariate Marshall–Olkin distribution can be characterized by the fact that each margin of any dimension satisfies the lack-of-memory property or by the following construction that we call *Marshall–Olkin machinery (MO-machinery)*.

Let $d > 2$, $\mathscr{P}_0 = \{S \subset \{1, \dots, d\} : S \neq \emptyset\}$, and $\{E_S\}_{S \in \mathscr{P}_0}$ be a collection of independent and exponentially distributed random variables with intensity $\lambda_S \geq 0$. Assume $\bar{\lambda}_j = \sum_{S: j \in S} \lambda_S > 0$, for $j = 1, \dots, d$. Then if

$$\tau_j = \min_{S: j \in S} E_S \tag{5.1}$$

the survival distribution of $(\tau_1, \dots, \tau_d)$ defines the *Multivariate Marshall–Olkin distribution (MMO)*

$$\bar{F}_{MMO}(\mathbf{t}) = \bar{F}_{\tau}(\mathbf{t}) = \exp\left(- \sum_{S \in \mathscr{P}_0} \lambda_S \max_{j \in S} t_j \right). \tag{5.2}$$

S. Mulinacci (✉)
Department of Statistics, University of Bologna, Bologna, Italy
e-mail: sabrina.mulinacci@unibo.it

U. Cherubini et al. (eds.), *Marshall–Olkin Distributions - Advances in Theory and Applications*, Springer Proceedings in Mathematics & Statistics 141,
DOI 10.1007/978-3-319-19039-6_5

The corresponding survival marginal distributions are

$$\bar{F}_{\tau_j}(t_j) = \exp(-\bar{\lambda}_j t_j)$$

and the corresponding survival copula (see Li [13]) is

$$\hat{C}_{MMO}(\mathbf{u}) = \prod_{S \in \mathscr{P}_0} \min_{k \in S} u_k^{\lambda_S / \bar{\lambda}_k}. \tag{5.3}$$

For an interpretation of the model, think of a system of d components $C_1, \ldots C_d$ (mechanical engines, electronic elements, credit obligors, life-insurance policy holders, etc.). The random variables E_S represent the arrival time of a shock causing the simultaneous failure (or default, or death, depending on the case) of those components C_j such that $j \in S$. So the time to failure of each component C_j is the random variable τ_j defined in (5.1).

Clearly, the MO-machinery can be applied to any family of random variables $(E_S)_{S \in \mathscr{P}_0}$ with support $[0, +\infty)$ not necessarily independent and exponentially distributed, in order to obtain Marshall–Olkin-like distributions and copula functions. The generalization to the case in which the E_S's are independent but not constrained to be exponentially distributed has already been widely considered in the literature and the most general formulation is provided by the Multivariate Marshall–Olkin distribution in Lin and Li [15].

It is, however, reasonable to consider concrete situations in which some dependence among the shocks causing the failure of the elements in the system considered exists. A natural way to introduce dependence is to consider some hidden common factor affecting all the shocks' arrival times. The idea to assume a common factor to induce dependence is at the basis of the original construction of the Archimedean copula function (see Marshall and Olkin [19]). In fact, it is well known that, if $\mathbf{T} = (T_1, \ldots, T_d)$ is a random vector with independent and exponentially distributed components and $\Lambda > 0$ is a positive and independent random variable with Laplace transform L, the random vector

$$\tau = \left(\frac{T_1}{\Lambda}, \ldots, \frac{T_d}{\Lambda} \right)$$

has the joint survival distribution

$$\bar{F}_\tau(t_1, \ldots, t_d) = E\left[\bar{F}_{\mathbf{T}}^{\Lambda}(t_1, \ldots, t_d) \right]$$

where $\bar{F}_{\mathbf{T}}$ is the joint survival distribution of the random vector $\mathbf{T}$, and the dependence structure given by an Archimedean copula with generator L. Hence the survival distribution of τ is obtained by power-mixing the survival distribution $\bar{F}_{\mathbf{T}}$ through the random variable Λ. The technique of constructing distributions through such a power-mixing technique was introduced in Marshall and Olkin [19] and then developed by Joe and Hu [12].

In this paper, we investigate the possibility to construct generalizations of the MMO distribution by inducing dependence among the shocks' arrival times through the power-mixing technique. In particular, we will see that the power-mixing technique is closed under the MO-machinery. More precisely, consider a random vector $\mathbf{E}$ with survival distribution $\bar{F}$ and another random vector $\mathbf{Z}$ whose survival distribution is obtained as a power-mixture of $\bar{F}$: if one applies the MO-machinery to both vectors, the survival distribution of the random vector generated from $\mathbf{Z}$ is the power-mixture of the one generated from $\mathbf{E}$. Finally, the power-mixing technique will be applied to construct generalizations of the Multivariate Generalized Marshall–Olkin distributions in order to allow for some dependence among the shocks' arrival times.

The paper is organized as follows.

In Sect. 5.2, the class of multivariate survival distributions to which the power-mixing technique can be applied is introduced and some relevant properties are pointed out: it is the class of Min-ID distributions introduced and studied in Joe and Hu [12] and Joe [11]. In Sect. 5.3, the closure property of the power-mixing technique under the MO-machinery is shown. Section 5.4 is devoted to a review about the generalized Marshall–Olkin distribution following Li and Pellerey [16] and Lin and Li [15]. In Sect. 5.5, an extension of the generalized Marshall–Olkin distribution is built in order to include dependence among the shocks affecting the system.

Throughout the paper, all vectors will be denoted with bold letters and equalities and inequalities among vectors will be meant componentwise. Moreover, by the sake of simplicity, all marginal survival distributions are assumed to be continuous and strictly decreasing.

5.2 Mixtures of Survival Distributions on $[0, +\infty)^d$

In this section, we will briefly recall the well-known concept of mixture of a survival min-infinitely divisible d-multivariate distribution with support $[0, +\infty)^d$ and we will present the main properties. We refer to Marshall and Olkin [19], Joe [10], Joe [11] and Joe and Hu [12] for a more detailed treatment.

5.2.1 *Min-ID Distributions*

Here, we summarize the main notions and properties of Min-Id distributions. We will consider only Min-Id distributions since we are interested in survival distribution functions and minima of random variables; the symmetric notion of Max-Id distributions can be trivially derived, replacing survival distribution functions with cumulative distribution functions and minima with maxima. All these concepts are well known and we refer the interested reader to Joe and Hu [12] and Joe [11] for a more complete presentation of this topic.

Definition 5.1 A *min-infinitely divisible multivariate distribution (Min-ID)* is a distribution F for which any positive power of the survival distribution function $\bar{F}$ is again a survival distribution function.

Even if positive powers of univariate survival distribution functions are always again survival distribution functions, for d-variate survival distribution functions, with $d \geq 2$, this is true for powers greater or equal to $d-1$.

In case of a Min-Id survival distribution $\bar{F}$, if $\mathbf{X} \sim \bar{F}^{1/n}$, for $n > 0$ and considering $\left(\mathbf{X}_i^{(n)}\right)_{i=1,\dots,n}$, n i.i.d. copies of $\mathbf{X}$, then

$$\left(\min_i X_{i,1}, \dots, \min_i X_{i,d}\right) \sim \bar{F}$$

and this fact justifies the name. Clearly Multivariate Extreme Value distributions with support $[0, +\infty)$ are Min-ID.

Remark 5.1 In Joe [11], several dependence properties of Min-ID distributions are analyzed. In particular, it is observed that the Min-ID requirement induces positive dependence; in fact, thanks to Theorems 2.3 and 2.6 in Joe [11], we have that any bivariate Min-ID distribution F is necessarily PQD (Positive Quadrant Dependent), that is

$$\hat{C}(u, v) \geq uv, \quad \forall (u, v) \in [0, 1]^2$$

where $\hat{C}$ is the survival copula associated to $\bar{F}$. As a consequence, a necessary condition for a multivariate distribution to be Min-ID is that all bivariate margins are PQD and this fact clearly implies that Min-ID is a strong positive dependence condition.

As it is well known, in case of a d-variate Extreme Value distribution $\bar{F}$, the above condition generalizes to the positive upper hortant dependence property

$$\hat{C}(u_1, \dots, u_d) \geq \prod_{j=1}^{d} u_j, \quad \forall \mathbf{u} \in [0, 1]^d$$

where $\hat{C}$ is the survival copula associated to $\bar{F}$.

Conditions for a multivariate distribution function to be Min-Id are given in Joe and Hu [12] and Joe [11].

An important subfamily of Min-ID distributions (strongly related to Multivariate Extreme Value distributions since they share the same copulas) is the class of *min-stable multivariate exponential distributions* that we now introduce.

Definition 5.2 Let $(\Omega, \mathscr{F}, \mathbb{P})$ be a probability space where a vector $\mathbf{T} = (T_1, \ldots, T_d)$ is defined with support $[0, +\infty)^d$. $\mathbf{T}$ is said to have a *min-stable multivariate exponential distribution (MSMVE)* if for all $k \in \{1, \ldots, d\}$, $1 \leq i_1 < \ldots < i_k \leq d$, $c_1, \ldots, c_k > 0$, $\min\{c_1 X_{i_1}, \ldots, c_k X_{i_k}\}$ is exponentially distributed.

We refer the reader to Joe [10] and, more recently, to Bernhart et al. [1] as further references on this topic.

The survival distribution function $\bar{F}$ of an MSMVE distribution is of type

$$\bar{F}_A(\mathbf{x}) = \exp\left(-A(\mathbf{x})\right), \ \mathbf{x} \in [0, +\infty)^d \tag{5.4}$$

with $A : [0, +\infty)^d \to [0, +\infty)$ homogeneous of order 1 (meaning that $A(\lambda \mathbf{x}) = \lambda A(\mathbf{x})$, for all $\lambda > 0$). If we standardize, assuming the marginals to be unit exponential, then $A(\mathbf{e_i}) = 1$, for all $\mathbf{e_i} = (0, \ldots, 0, 1, 0, \ldots, 0)$: A is called a *stable tail dependence function* (see Joe [10] and Theorem 6.2 in Joe [11]).
The survival copula of an MSMVE-distributed random variable is of type

$$\hat{C}_A(\mathbf{u}) = \exp\left(-A(-\ln u_1, \ldots, -\ln u_d)\right), \ \forall \mathbf{u} \in [0, 1]^d \tag{5.5}$$

It is a well-known fact (see, for example, Sect. 6.2 in Joe [11]) that the family of copulas of type (5.5) coincides with the family of extreme value multivariate copula functions.

The MSMVE family is a subclass of Min-ID distributions. In fact, if $\mathbf{T}$ is a MSMVE distributed d-vector with survival distribution function $\bar{F}_A$ for a given A (see (5.4)), then if $\lambda > 0$, the d-vector $\mathbf{T}_\lambda = \left(\frac{T_1}{\lambda}, \ldots, \frac{T_d}{\lambda}\right)$ has the survival distribution

$$\bar{F}_{\mathbf{T}_\lambda}(\mathbf{t}) = \bar{F}_A(\lambda \mathbf{t}) = \exp\left(-A(\lambda \mathbf{t})\right) = \bar{F}_A^\lambda(\mathbf{t}).$$

Remark 5.2 In Joe and Hu [12] it is proved that if one considers survival Multivariate Extreme Value distributions with all the marginals in the same family, then the closure property under the minimum characterizing MSMVE distributions is preserved: this is not the case for general Min-ID distributions.

Starting from the survival copula (5.5) of an MSMVE-distribution, other Min-ID distributions can be recovered. In fact, applying Sklar's theorem, we can allow for different marginal survival distribution functions $\bar{G}_1, \ldots, \bar{G}_d$ all of them having $[0, +\infty)$ as support. If we consider the cumulative hazard functions $H_i(x) = -\ln \bar{G}_i(x)$, $i = 1, \ldots, d$, we get the survival distribution functions with support $[0, +\infty)^d$ of type

$$\begin{aligned} \bar{G}_A(\mathbf{x}) &= \bar{G}_A(\mathbf{x}; H_1, \ldots, H_d) = \\ &= \hat{C}_A\left(\bar{G}_1(x_1), \ldots, \bar{G}_d(x_d)\right) = \\ &= \exp\left(-A\left(-\ln \bar{G}_1(x_1), \ldots, -\ln \bar{G}_d(x_d)\right)\right) = \\ &= \bar{F}_A\left(H_1(x_1), \ldots, H_d(x_d)\right). \end{aligned} \tag{5.6}$$

G_A is again Min-ID. In fact, for all $\lambda > 0$, $\bar{G}_A^\lambda$ is the survival distribution of the d-vector $\left(H_1^{-1}\left(\frac{T_1}{\lambda}\right), \ldots, H_d^{-1}\left(\frac{T_d}{\lambda}\right)\right)$, where $\mathbf{T} = (T_1, \ldots, T_d)$ is MSMVE with standard unit exponential marginals.

5.2.2 Power-Mixtures of Survival Min-ID Distributions

Let us consider a survival Min-ID d-dimensional distribution function $\bar{F}$ and a positive random variable Λ with Laplace transform L. The corresponding *mixed survival distribution* is defined as

$$\bar{F}_\Lambda(\mathbf{z}) = E\left[\bar{F}^\Lambda(\mathbf{z})\right] = L\left(-\ln \bar{F}(\mathbf{z})\right). \tag{5.7}$$

If $\hat{C}$ is the survival copula associated to $\bar{F}$, then the survival copula associated to the mixture (5.7) is

$$\hat{C}_\Lambda(\mathbf{u}) = L\left(-\ln \hat{C}\left(e^{-L^{-1}(u_1)}, \ldots, e^{-L^{-1}(u_d)}\right)\right). \tag{5.8}$$

Remark 5.3 Notice that $\bar{F}_\Lambda$ $(\hat{C}_\Lambda)$ is exchangeable if and only if $\bar{F}$ $(\hat{C})$ is.

Remark 5.4 The power-mixing technique preserves the lower orthant ordering. In fact, if $\bar{F}_1(\mathbf{z}) \geq \bar{F}_2(\mathbf{z})$, for all $z \in [0, +\infty)^d$, then $\bar{F}_{1,\Lambda}(\mathbf{z}) \geq \bar{F}_{2,\Lambda}(\mathbf{z})$, for all $z \in [0, +\infty)^d$.

In Joe and Hu [12], some subfamilies of such distributions are introduced and their dependence properties analyzed.

As particular cases, here we consider mixtures of survival distributions of type (5.4) and (5.6). Let Λ be a positive random variable and L its Laplace transform. The mixed survival distributions corresponding to (5.4) and (5.6), respectively, are

$$\begin{aligned} \bar{F}_{A,\Lambda}(\mathbf{z}) &= E\left[\bar{F}_A^\Lambda(\mathbf{z})\right] = \\ &= L\left(-\ln\left(\bar{F}_A(\mathbf{z})\right)\right) = \\ &= L \circ A(\mathbf{z}) \end{aligned} \tag{5.9}$$

and

$$\begin{aligned} \bar{G}_{A,\Lambda}(\mathbf{z}; H_1, \ldots, H_d) &= E\left[\bar{G}_A^\Lambda(\mathbf{z}; H_1, \ldots, H_d)\right] = \\ &= L\left(-\ln\left(\bar{G}_A(\mathbf{z}; H_1, \ldots, H_d)\right)\right) = \\ &= L\left(-\ln\left(\bar{F}_A\left(H_1(z_1), \ldots, H_d(z_d)\right)\right)\right) = \\ &= L \circ A\left(H_1(z_1), \ldots, H_d(z_d)\right). \end{aligned} \tag{5.10}$$

In particular, it follows that

$$\bar{G}_{A,\Lambda}(\mathbf{z}; H_1, \ldots, H_d) = \bar{F}_{A,\Lambda}(H_1(z_1), \ldots, H_d(z_d)).$$

Notice that $\bar{G}_{A,\Lambda}(\mathbf{z}; H_1, \ldots, H_d)$ is the survival distribution function of the vector

$$\mathbf{Z} = \left(H_1^{-1}\left(\frac{T_1}{\Lambda}\right), \ldots, H_d^{-1}\left(\frac{T_d}{\Lambda}\right)\right)$$

where $\mathbf{T} = (T_1, \ldots, T_d)$ has survival distribution $\bar{F}_A$ and the positive random variable Λ is independent of the d-vector $\mathbf{T}$. Since $\bar{F}_{A,\Lambda}(\mathbf{z})$ is the survival distribution function of $\mathbf{W} = \left(\frac{T_1}{\Lambda}, \ldots, \frac{T_d}{\Lambda}\right)$, (5.9) and (5.10) share the same survival copula function

$$\hat{C}_{A,\Lambda}(\mathbf{u}) = L\left(-\ln \hat{C}_A\left(e^{-L^{-1}(u_1)}, \ldots, e^{-L^{-1}(u_d)}\right)\right). \tag{5.11}$$

Remark 5.5 Copulas of type (5.11) are of Archimax type (see Charpentier et al. [4], for recent results on multivariate Archimax copulas). Multivariate Archimax copulas are of type

$$C_{\psi,D}(\mathbf{u}) = \psi\left(-\ln D\left(e^{-\psi^{-1}(u_1)}, \ldots, e^{-\psi^{-1}(u_d)}\right)\right)$$

where D is a d-variate extreme value copula and ψ is the generator of a d-variate Archimedean copula (see (3) in Charpentier et al. [4]). Since (see (5.5)) $\hat{C}_A$ is an extreme value copula, copulas of type (5.11) coincide with the subset of Archimax copulas with a completely monotone generator (see Sect. 3.1 in Charpentier et al. [4]).

5.2.3 Conditional Mixed Survival Min-ID Distributions

In this subsection we will analyze the conditional distribution of a nonnegative d-random vector whose survival distribution is a mixture of a survival Min-Id distribution function.

More precisely, let $\bar{F}_\Lambda$ be defined as in (5.7) and let $\mathbf{Z}$ be a d-random vector accordingly distributed. We are interested in studying, for $\mathbf{s} \in [0, +\infty)^d$

$$\bar{F}_{\Lambda,\mathbf{s}}(\mathbf{t}) = \mathbb{P}(\mathbf{Z} > \mathbf{t} + \mathbf{s} | \mathbf{Z} > \mathbf{s}) \ \forall \mathbf{t} \in [0, +\infty)^d.$$

We start considering the original unmixed Min-ID survival distribution function $\bar{F}$ (see (5.7)) and a d-random vector $\mathbf{X} = (X_1, \ldots, X_d)$ accordingly distributed. We set

$$\bar{F}_\mathbf{s}(\mathbf{t}) = \mathbb{P}(\mathbf{X} > \mathbf{t} + \mathbf{s} | \mathbf{X} > \mathbf{s}).$$

$\bar{F}_\mathbf{s}$ is clearly again Min-ID since $\bar{F}_\mathbf{s}^\lambda(\mathbf{t}) = \frac{\bar{F}^\lambda(\mathbf{t}+\mathbf{s})}{\bar{F}^\lambda(\mathbf{s})}$, for any $\lambda > 0$.

Proposition 5.1 *$\bar{F}_{\Lambda,\mathbf{s}}$ is the power-mixture of $\bar{F}_{\mathbf{s}}$. More precisely,*

$$\bar{F}_{\Lambda,\mathbf{s}}(\mathbf{t}) = L_{\mathbf{s}}\left(-\ln \bar{F}_{\mathbf{s}}(\mathbf{t})\right) \tag{5.12}$$

where $L_{\mathbf{s}}$ is the Laplace transform of the positive random variable $\Lambda_{\mathbf{s}}$ with cumulative distribution function

$$F_{\Lambda_{\mathbf{s}}}(y) = \frac{1}{L\left(-\ln \bar{F}(\mathbf{s})\right)} \int_0^y \bar{F}^w(\mathbf{s}) dF_\Lambda(w),\ y \in [0, +\infty)$$

where F_Λ and L are, respectively, the cumulative distribution function and the Laplace transform of Λ.

Proof

$$\bar{F}_{\Lambda,\mathbf{s}}(\mathbf{t}) = \frac{\bar{F}_\Lambda(\mathbf{t}+\mathbf{s})}{\bar{F}_\Lambda(\mathbf{s})} = \frac{L\left(-\ln \bar{F}(\mathbf{t}+\mathbf{s})\right)}{L\left(-\ln \bar{F}(\mathbf{s})\right)} = \frac{L\left(-\ln \bar{F}(\mathbf{s}) - \ln \frac{\bar{F}(\mathbf{t}+\mathbf{s})}{\bar{F}(\mathbf{s})}\right)}{L\left(-\ln \bar{F}(\mathbf{s})\right)}.$$

Set

$$L_{\mathbf{s}}(x) = \frac{L\left(-\ln \bar{F}(\mathbf{s}) + x\right)}{L\left(-\ln \bar{F}(\mathbf{s})\right)}.$$

It can be easily verified that $L_{\mathbf{s}}$ is the Laplace transform of the positive random variable $\Lambda_{\mathbf{s}}$ with cumulative distribution function

$$F_{\Lambda_{\mathbf{s}}}(y) = \frac{1}{L\left(-\ln \bar{F}(\mathbf{s})\right)} \int_0^y \bar{F}^w(\mathbf{s}) dF_\Lambda(w),\ y \in [0, +\infty).$$

Hence

$$\bar{F}_{\Lambda,\mathbf{s}}(\mathbf{t}) = L_{\mathbf{s}}\left(-\ln \bar{F}_{\mathbf{s}}(\mathbf{t})\right).$$

By (5.12), for every $\mathbf{s} \in [0, +\infty)^d$, the conditional survival distribution $\bar{F}_{\Lambda,\mathbf{s}}$ is again a mixture of the survival Min-ID conditional distribution $\bar{F}_{\mathbf{s}}$ through a suitable positive random variable $\Lambda_{\mathbf{s}}$. As a consequence, if $\hat{C}_{\mathbf{s}}$ is the survival copula associated to $\bar{F}_{\mathbf{s}}$, then

$$\hat{C}_{\Lambda,\mathbf{s}}(\mathbf{u}) = L_{\mathbf{s}}\left(-\ln \hat{C}_{\mathbf{s}}\left(e^{-L_{\mathbf{s}}^{-1}(u_1)}, \dots, e^{-L_{\mathbf{s}}^{-1}(u_d)}\right)\right)$$

is the survival copula associated to $\bar{F}_{\Lambda,\mathbf{s}}$ (see also [3]).

5.3 Power-Mixture Closure Under the MO-machinery

In this section, we will show that, starting from a survival multivariate distribution that is the mixture of a survival Min-ID through a positive random variable Λ and applying the MO-machinery, then the obtained survival distribution is again the mixture through the same Λ of the survival distribution obtained by applying the MO-machinery to the unmixed original survival Min-ID distribution. In particular, we will show that the Min-ID property is closed under the MO-machinery.

5.3.1 The General MO-machinery

In this subsection, we formalize a general version of the MO-machinery model in order to allow for shocks whose arrival times are not necessarily independent and exponentially distributed.

Let $d > 2$, $\mathscr{P}_0 = \{S \subset \{1, \dots, d\} : S \neq \emptyset\}$ and $\pi : \mathscr{P}_0 \to \{1, \dots, 2^d - 1\}$ be some ordering on $\mathscr{P}_0$. We consider a random vector $\mathbf{E} = (E_1, \dots, E_{2^d-1})$ whose components have support $[0, +\infty)$ or $\{+\infty\}$. We denote with $\bar{F}_{\mathbf{E}}$ the corresponding survival distribution function.

The idea behind the above model is the following: let us consider a system with d components $C_1, \dots C_d$; each E_i is the time-arrival of a shock causing the simultaneous default of those components C_j such that $j \in \pi^{-1}(i)$: if $E_i \equiv +\infty$, then there is no common shock involving simultaneously all the components C_j with $j \in \pi^{-1}(i)$.

Assumption In order to exclude the case of a component with perpetual life, we assume that for every $j \in \{1, \dots, d\}$, there exists at least one i with $j \in \pi^{-1}(i)$ such that E_i has the whole positive half-line as support.

If

$$M_j = \min_{i: j \in \pi^{-1}(i)} E_i, \ j = 1, \dots, d,$$

the survival distribution function of the random vector $\mathbf{M} = (M_1, \dots, M_d)$ is

$$\begin{aligned} \bar{F}_{\mathbf{M}}(\mathbf{t}) &= \mathbb{P}\left(M_j > t_j, j = 1, \dots, d\right) = \\ &= \mathbb{P}\left(\min\left(E_i : j \in \pi^{-1}(i)\right) > t_j, j = 1, \dots, d\right) = \\ &= \mathbb{P}\left(E_i > \max(t_j : j \in \pi^{-1}(i)), i = 1, \dots, 2^d - 1\right) = \\ &= \bar{F}_{\mathbf{E}}(s_1, \dots, s_{2^d-1}) \end{aligned} \tag{5.13}$$

where $s_i = \max\left(t_j : j \in \pi^{-1}(i)\right)$.

The marginal survival distributions are

$$\bar{F}_{M_j}(t) = \mathbb{P}\left(E_i > t, \forall i : j \in \pi^{-1}(i)\right) = \exp\left(-W_j(t)\right).$$

and the associated survival copula is

$$\hat{C}_{\mathbf{M}}(\mathbf{u}) = \bar{F}_{\mathbf{E}}\left(\max_{j\in\pi^{-1}(1)}\left(W_j^{-1}(-\ln u_j)\right), \ldots, \max_{j\in\pi^{-1}(2^d-1)}\left(W_j^{-1}(-\ln u_j)\right)\right) \tag{5.14}$$

Remark 5.6 The MO-machinery considered in this paper is a particular specification of the more general concept of MO-machinery considered in Durante et al. [6] and of the generalizations of the Marshall–Olkin model presented in Frostig and Pellerey [9]: here the effect of the random variables affecting the system (or portions of it) is given by the minimum of their time arrivals.

Example 5.1 If $\bar{F}_{\mathbf{E}}$ is of type (5.6) for some homogeneous $A : [0, +\infty)^{2^d-1} \to [0, +\infty)$, then

$$\begin{aligned}\bar{F}_{\mathbf{M}}(\mathbf{t}) &= \bar{F}_A(H_1(s_1), \ldots, H_{2^d-1}(s_{2^d-1})) = \\ &= \exp\left(-A\left(\max_{j\in\pi^{-1}(1)} H_1(t_j), \ldots, \max_{j\in\pi^{-1}(2^d-1)} H_{2^d-1}(t_j)\right)\right)\end{aligned}$$

and

$$\bar{F}_{M_j}(t) = \exp\left(-K_j(t)\right)$$

with $K_j(t) = A\left(\mathbf{h}_j(t)\right)$ where $\mathbf{h}_j(t)$ is a $2^d - 1$-vector whose ith component is $H_i(t)$ if $j \in \pi^{-1}(i)$ and 0 otherwise. The associated survival copula is

$$\hat{C}_{\mathbf{M}}(\mathbf{u}) = e^{-A\left(\max_{j\in\pi^{-1}(1)}\left(H_1 \circ K_j^{-1}(-\ln u_j)\right), \ldots, \max_{j\in\pi^{-1}(2^d-1)}\left(H_{2^d-1} \circ K_j^{-1}(-\ln u_j)\right)\right)}.$$

Example 5.2 **The bivariate case.** Let $d = 2$, $\pi(\{1\}) = 1$, $\pi(\{2\}) = 2$, and $\pi(\{1,2\}) = 3$ and $\mathbf{E} = (E_1, E_2, E_3)$. Then $\mathbf{M} = (M_1, M_2)$ with $M_1 = \min(E_1, E_3)$ and $M_2 = \min(E_2, E_3)$.
We have that

$$\bar{F}_{\mathbf{M}}(t_1, t_2) = \bar{F}_{\mathbf{E}}(t_1, t_2, \max(t_1, t_2)),$$

the marginal survival distributions are

$$\bar{F}_{M_j}(t) = \bar{F}_{E_j, E_3}(t, t)$$

and, if $W_j(t) = -\ln \bar{F}_{E_j,E_3}(t)$, the associated copula function is

$$\hat{C}_{\mathbf{M}}(\mathbf{u}) = \bar{F}_{\mathbf{E}}\left(W_1^{-1}(-\ln u_1), W_2^{-1}(-\ln u_2), \max\left(W_1^{-1}(-\ln u_1), W_2^{-1}(-\ln u_2)\right)\right).$$

If, in particular, the survival distribution $\bar{F}_{\mathbf{E}}$ is of type (5.6), we have

$$\bar{F}_{\mathbf{M}}(t_1, t_2) = \exp\left(-A\left(H_1(t_1), H_2(t_2), H_3(\max(t_1, t_2))\right)\right)$$

with marginal survival distributions

$$\bar{F}_{M_j}(t) = \exp\left(-K_j(t)\right), \ j = 1, 2$$

where $K_1(x) = A\left(H_1(x), 0, H_3(x)\right)$ and $K_2(x) = A\left(H_2(x), 0, H_3(x)\right)$, and survival copula

$$\hat{C}_{\mathbf{M}}(\mathbf{u}) = e^{-A\left(H_1 \circ K_1^{-1}(-\ln u_1), H_2 \circ K_2^{-1}(-\ln u_2), \max\left(H_3 \circ K_1^{-1}(-\ln u_1), H_3 \circ K_2^{-1}(-\ln u_2)\right)\right)}.$$

5.3.2 *Min-ID Closure Under the MO-machinery*

Proposition 5.2 *If $F_{\mathbf{E}}$ is Min-ID, then $F_{\mathbf{M}}$ is Min-ID as well.*

Proof Given $\lambda > 0$, let us consider a probability space $(\Omega, \mathscr{F}, \mathbb{P})$ and a $2^d - 1$-vector $\mathbf{E}_\lambda$ whose survival distribution is $\bar{F}_{\mathbf{E}}^{\lambda}$. Applying to $\mathbf{E}_\lambda$ the MO-machinery described in the previous subsection, we have that

$$W(t_1, \ldots, t_d) - \bar{F}_{\mathbf{E}}^{\lambda}(s_1, \ldots, s_{2^d-1})$$

(with $s_i = \max\left(t_j : j \in \pi^{-1}(i)\right)$) is a d-variate survival distribution function and (see (5.13))

$$W(t_1, \ldots, t_d) = \bar{F}_M^{\lambda}(t_1, \ldots, t_d).$$

Hence $F_{\mathbf{M}}$ is again Min-ID.

As we saw in Sect. 5.2, MSMVE distributions are Min-ID and this fact is a consequence of the closure property under the min operator. Similarly, the closure property under the min operator is preserved by the MO-machinery.

Proposition 5.3 *If $F_{\mathbf{E}}$ is MSMVE, then $F_{\mathbf{M}}$ is MSMVE as well.*

Proof $\bar{F}_{\mathbf{M}}$ is the survival distribution of $\mathbf{M} = (M_1, \ldots, M_d)$ with

$$M_j = \min\left(E_i : j \in \pi^{-1}(i)\right), \ j = 1, \ldots, d,$$

and, if $k \in \{1, \ldots, d\}$ and $1 \leq i_1 < i_2 < \cdots < i_k \leq d$ and $c_{i_j} > 0$ for $j = 1, \ldots, k$

$$\min\{c_{i_1} M_{i_1}, \ldots, c_{i_k} M_{i_k}\} = \min_{j \in J}\{d_j E_j\}$$

where $J = \{i : \exists \rho \in \{1, \ldots, k\} : i_\rho \in \pi^{-1}(i)\}$ and $d_j = \min\{c_{i_\rho} : \rho \in \{1, \ldots, k\},\ i_\rho \in \pi^{-1}(j)\}$. It follows that $\min_{j \in J}\{d_j E_j\}$ is again exponentially distributed being $F_{\mathbf{E}}$ MSMVE. Hence $F_{\mathbf{M}}$ is MSMVE.

Remark 5.7 Notice that if $F_{\mathbf{E}}$ is MSMVE, then $\bar{F}_{\mathbf{M}}^{\lambda}$ is the survival distribution of $\left(\frac{M_1}{\lambda}, \ldots, \frac{M_d}{\lambda}\right)$.

5.3.3 Power-Mixture Closure Under the MO-machinery

Let us assume that $F_{\mathbf{E}}$ is Min-ID and let $\bar{F}_\Lambda$ be the mixed transform of $\bar{F}_{\mathbf{E}}$ through $\Lambda > 0$ with Laplace transform L. Let $(\Omega, \mathscr{F}, \mathbb{P})$ be a probability space and $\mathbf{Z} = (Z_1, \ldots, Z_{2^d-1})$ be a random vector distributed according to $\bar{F}_\Lambda$.

Proposition 5.4 *The mixture of the survival Min-ID distribution $\bar{F}_M$ is the survival distribution of the vector $\tau = (\tau_1, \ldots, \tau_d)$ with*

$$\tau_k = \min_{i : k \in \pi^{-1}(i)} Z_i, \qquad k = 1, \ldots, d.$$

Proof Setting again $s_i = \max_{k \in \pi^{-1}(i)} t_k$,

$$\begin{aligned}
\bar{F}_\tau(\mathbf{t}) &= \mathbb{P}\left(\tau_k > t_k : k = 1, \ldots, d\right) = \\
&= \mathbb{P}\left(\min\left(Z_i : k \in \pi^{-1}(i)\right) > t_k, k = 1, \ldots, d\right) = \\
&= \mathbb{P}\left(Z_i > \max\left(t_k : k \in \pi^{-1}(i)\right), i = 1, \ldots, 2^d - 1\right) = \\
&= \bar{F}_\Lambda(s_1, \ldots, s_{2^d-1}) = \\
&= E\left[\bar{F}_{\mathbf{E}}^{\Lambda}\left(s_1, \ldots, s_{2^d-1}\right)\right] = \\
&= E\left[\bar{F}_{\mathbf{M}}^{\Lambda}(\mathbf{t})\right]
\end{aligned}$$

that is, $\bar{F}_\tau$ is the mixture of $\bar{F}_{\mathbf{M}}$ through Λ and

$$\bar{F}_\tau(\mathbf{t}) = L\left(-\ln \bar{F}_{\mathbf{M}}(\mathbf{t})\right). \tag{5.15}$$

As a consequence, the associated survival copula function $\hat{C}_\tau$ is

$$\hat{C}_\tau(\mathbf{u}) = L\left(-\ln \hat{C}_{\mathbf{M}}\left(e^{-L^{-1}(u_1)}, \ldots, e^{-L^{-1}(u_d)}\right)\right) \tag{5.16}$$

where $\hat{C}_{\mathbf{M}}$ is the survival copula function associated to $\bar{F}_{\mathbf{M}}$ given in (5.14).

Example 5.3 **The scale-mixtures of Marshall–Olkin distributions**
In Li [14] the scale-mixing technique is applied to multivariate extreme value distributions and the tail behavior of the obtained distributions is analyzed. In particular, the scale-mixture of the Marshall–Olkin distribution (SMMO distribution) is obtained from the multivariate Marshall–Olkin distribution. These distributions and copulas have also been considered in Bernhart et al. [2] and in Mai et al. [17].

SMMO distributions represent a particular example of the construction presented in this section. In fact, the multidimensional Marshall–Olkin distribution is of type (5.13) with the underlying random vector $\mathbf{E}$ having independent and exponentially distributed components: since obviously $F_{\mathbf{E}}$ is a MSMVE distribution, according to Remark 5.7, power-mixing is equivalent to scale mixing.

More precisely, the SMMO distribution and the SMMO copula can be recovered from (5.15) and (5.2) and from (5.16) and (5.3), respectively. In fact, since here $\mathbf{E}$ has independent components, $\bar{F}_{\mathbf{E}}$ is the product of the survival marginal distributions and so there is no need of the ordering function π on $\mathscr{P}_0$: hence we can simplify the notation denoting with E_S the random variable E_i such that $\pi(S) = i$. The formulas we get are:

$$\bar{F}_{SMMO}(\mathbf{t}) = L\left(-\ln \bar{F}_{MMO}(\mathbf{t})\right) = L\left(\sum_{S\in\mathscr{P}_0} \lambda_S \max_{j\in S} t_j\right)$$

and

$$\hat{C}_{SMMO}(\mathbf{u}) = L\left(-\ln \hat{C}_{MO}\left(e^{-L^{-1}(u_1)}, \ldots, e^{-L^{-1}(u_d)}\right)\right) =$$
$$= L\left(\sum_{S\in\mathscr{P}_0} \max_{k\in S}\left(L^{-1}(u_k)\frac{\lambda_S}{\sum_{V\in\mathscr{P}_0:k\in V}\lambda_V}\right)\right).$$

5.4 The Generalized Marshall–Olkin Type Distributions

In this section, we describe the *Generalized Marshall–Olkin distributions* and the *Generalized Marshall–Olkin copula functions* introduced in the bivariate case in Li and Pellerey [16] and in the multivariate case on Lin and Li [15]. These distributions can be obtained by applying the MO-machinery to a random vector $\mathbf{E}$ whose components, even if again independent, are not constrained to be exponentially distributed. The distribution of $\mathbf{E}$ is Min-ID and so all the results of Sect. 5.3 apply.

More precisely, let us consider the setting of Example 5.1 which $A(\mathbf{x}) = x_1 + \cdots + x_{2^d-1}$. Since such a function A is exchangeable, we can simplify the notation, being the induced survival distribution independent of the ordering function π. Hence we will denote the elements of $\mathbf{E}$ with E_S for $S \in \mathscr{P}_0$.

So, following Example 5.1, we get the *Multivariate Generalized Marshall–Olkin distribution with generating functions* (MGMO) $(H_S)_{S\in\mathscr{P}_0}$

$$\bar{F}_{MGMO}(\mathbf{t}) = \bar{F}_{MGMO}\left(\mathbf{t};\,(H_S)_{S\in\mathscr{P}_0}\right) = \exp\left(-\sum_{S\in\mathscr{P}_0}\max_{j\in S} H_S(t_j)\right). \tag{5.17}$$

The jth marginal survival distribution is

$$\bar{F}_{MGMO,j}(t) = \exp\left(-K_j(t)\right)$$

with

$$K_j(t) = \sum_{S:j\in S} H_S(t)$$

and the associated survival copula is

$$\begin{aligned}\hat{C}_{MGMO}(\mathbf{u}) &= \hat{C}_{MGMO}\left(\mathbf{u};\,(H_S)_{S\in\mathscr{P}_0}\right) = \\ &= \exp\left(-\sum_{S\in\mathscr{P}_0}\max_{j\in S} H_S\circ K_j^{-1}(-\ln u_j)\right)\end{aligned} \tag{5.18}$$

Thanks to Proposition 5.2, the MGMO distribution is Min-ID.

Example 5.4 **The bivariate case: Li and Pellerey** [16]
Let us consider $d = 2$ and $E_1 = E_{\{1\}}$, $E_1 = E_{\{2\}}$ and $E_3 = E_{\{1,2\}}$. Then, from (5.17) we get the *Generalized Marshall–Olkin distribution with generating functions* $(H_i)_{i=1,2,3}$

$$\begin{aligned}\bar{F}_{GMO}(x_1,x_2) &= \bar{F}_{GMO}(x_1,x_2;\,(H_i)_{i=1,2,3})(x_1,x_2) = \\ &= \exp(-H_1(x_1) - H_2(x_2) - H_3(\max(x_1,x_2))).\end{aligned}$$

If $K_i = H_i + H_3$, for $i = 1, 2$, then the corresponding marginal survival distributions are

$$\bar{F}_{GMO,j}(x) = \exp(-K_j(x)),\ j = 1, 2$$

while the associated survival copula function is, by (5.18),

$$\begin{aligned}\hat{C}_{GMO}(u,v) &= \hat{C}_{GMO}(x_1,x_2;\,(H_i)_{i=1,2,3})(u,v) = \\ &= \exp\Big(-H_1\circ K_1^{-1}(-\ln u) - H_2\circ K_2^{-1}(-\ln v) + \\ &\quad -H_3\left(\max(K_1^{-1}(-\ln u),\,K_2^{-1}(-\ln v))\right)\Big) =\end{aligned}$$

$$= uv\exp\left(H_3 \circ K_1^{-1}(-\ln u) + H_3 \circ K_2^{-1}(-\ln v) + \right.$$
$$\left. -H_3\left(\max(K_1^{-1}(-\ln u), K_2^{-1}(-\ln v))\right)\right) =$$
$$= uv\exp\left(H_3\left(\min\left(K_1^{-1}(-\ln u), K_2^{-1}(-\ln v)\right)\right)\right).$$

Since, thanks to Proposition 5.2, the Generalized Marshall–Olkin distributions are Min-ID, they are PQD (see Remark 5.1), as stated in Proposition 2.1 in Li and Pellerey [16].

Example 5.5 If $H_S(x) = \lambda_S H(x)$ for $\lambda_S \geq 0$, we recover from (5.17) the Muliere and Scarsini distribution (see Muliere and Scarsini [21])

$$\bar{F}_{MS}(\mathbf{t}) = \exp\left(-\sum_{S \in \mathscr{P}_0} \max_{j \in S} \lambda_S H(t_j)\right)$$

and from (5.18) the multivariate Marshall–Olkin copula in (5.3).

In Lin and Li [15], several results are proved that generalize those shown in Li and Pellerey [16] in the bivariate case. In particular, it is shown that

- an MGMO-distributed d-vector M is always *positively associated*, that is

$$Cov\left(\phi(M_1, \ldots, M_d), \psi(M_1, \ldots, M_d)\right) \geq 0$$

  for any increasing functions ϕ and ψ with finite covariance;
- an MGMO copula is *positively upper orthant dependent*, that is

$$\hat{C}_{MGMO}(\mathbf{u}) \geq \prod_{i=1}^{d} u_i, \ \mathbf{u} \in [0, 1]^d. \tag{5.19}$$

Moreover, the authors present some comparison results among different MGMO distributions and among different MGMO copulas.

More precisely, let us consider two d-dimensional MGMO distributions with generating functions $\{H_S : S \in \mathscr{P}_0\}$ and $\{V_S : S \in \mathscr{P}_0\}$, respectively. Some sufficient conditions are given to guarantee the dominance relation

$$\hat{C}_{MGMO}\left(\mathbf{u}; (H_j)_{j=1,\ldots,2^d-1}\right) \geq \hat{C}_{MGMO}\left(\mathbf{u}; (V_j)_{j=1,\ldots,2^d-1}\right) \tag{5.20}$$

for all $\mathbf{u} \in [0, 1]^d$.

In particular (5.20) is satisfied if one of the following conditions holds:

1. **C1**:
$$H_I \circ \left(\sum_{S:i\in S} H_S \right)^{-1}(x) \geq V_I \circ \left(\sum_{S:i\in S} V_S \right)^{-1}(x),\ \forall x \geq 0$$
for all $I \in \mathscr{P}_0$ such that $|S| \geq 2$[1] and for all $i \in S$ (see Theorem 4 in Lin and Li [15]);
2. **C2**: the two d-dimensional MGMO distributions have common univariate margins and $H_S \geq V_S$ for all $S \in \mathscr{P}_0$ with $|S| \geq 2$ (see Corollary 2 in Lin and Li [15]);
3. **C3**:
   a.
   $$H_S \circ H_I^{-1}(x) = V_S \circ V_I^{-1}(x),\ \forall x \geq 0$$
   for all $S, I \in \mathscr{P}_0$ such that $|S| \geq 2$, $|I| \geq 2$ and $S \cap I \neq \emptyset$ and
   b. for all $S \in \mathscr{P}_0$, with $|S| \geq 2$, and for all $i \in S$,
   $$H_{\{i\}} \circ H_S^{-1}(x) \leq V_{\{i\}} \circ V_S^{-1}(x),\ \forall x \geq 0$$
   (see Proposition 8 in Lin and Li [15]).

In Frostig and Pellerey [9] other conditions for (5.20) are provided. In particular, it is straightforward to write sufficient conditions on the generating functions so that their Corollary 3.1 applies.

In Sect. 4 of Lin and Li [15] the residual life

$$\mathbf{X}_t = [X_1 - t, \ldots, X_d - t | X_1 > t, \ldots, X_d > t],\ t > 0$$

of a vector $\mathbf{X}$ distributed according to an MGMO distribution is considered and the dynamic behavior of the corresponding survival copula is studied.

The survival distribution of the residual life is

$$\bar{F}_{\mathbf{X}_t}(\mathbf{t}) = \exp\left(-\sum_{S\in\mathscr{P}_0} W_{S,t}\left(\max_{i\in S} t_i \right) \right) \tag{5.21}$$

where

$$W_{S,t}(x) = H_S(x+t) - H_S(t)$$

[1] Here and in the sequel, given a finite set A, with $|A|$ we denote the number of elements in A.

while the associated copula is (see Proposition 4 in Lin and Li [15])

$$\hat{C}_{\mathbf{X}_t}(\mathbf{u}) = \exp\left(- \sum_{S \in \mathscr{P}_0} \max_{i \in S} W_{S,t} \circ \left(\sum_{I : i \in I} W_{I,t} \right)^{-1} (-\ln u_i) \right) \tag{5.22}$$

(5.21) and (5.22) generalize the bivariate analogous ones in Sect. 4 of Li and Pellerey [16].

5.5 The Mixed Generalized Marshall–Olkin Distribution

Since MGMO distributions are Min-ID, we can consider power-mixtures of their survival versions. More precisely, in this section, we will apply (5.7) and (5.8) to obtain new distributions and copula functions.

More precisely, if $\Lambda > 0$ is a positive random variable with Laplace transform L, we call *Mixed Multidimensional Generalized Marshall–Olkin distribution (Mix-MGMO) with generating functions* $(H_S)_{S \in \mathscr{P}_0}$ the distribution

$$\begin{aligned} \bar{F}_{MGMO,\Lambda}(\mathbf{t}) &= \bar{F}_{MGMO,\Lambda}\left(\mathbf{t}; (H_S)_{S \in \mathscr{P}_0}\right) = \\ &= L\left(-\ln \bar{F}_{MGMO}(\mathbf{t})\right) = \\ &= L\left(\sum_{S \in \mathscr{P}_0} \max_{j \in S} H_S(t_j) \right). \end{aligned} \tag{5.23}$$

We recall that, as stated in Proposition 5.4, this is the distribution of the random vector $\tau = (\tau_1, \ldots, \tau_d)$ such that $\tau_i = \min_{S : i \in S} Z_S$ where $(Z_S)_{S \in \mathscr{P}_0}$ is distributed according to the survival distribution function

$$\bar{F}_\Lambda(\mathbf{x}) = L\left(\sum_{S \in \mathscr{P}_0} H_S(x_S) \right). \tag{5.24}$$

where $\mathbf{x} = (x_S)_{S \in \mathscr{P}_0}$.

Equality (5.24) can be rewritten as

$$\bar{F}_\Lambda(\mathbf{x}) = L\left(\sum_{S \in \mathscr{P}_0} L^{-1}(\bar{F}_S(x_S)) \right)$$

where $\bar{F}_S(x) = L \circ H_S(x)$, meaning that, thanks to Sklar's theorem, the dependence structure of the family of random variables $(Z_S)_{S \in \mathscr{P}_0}$ is of Archimedean type with generator L. Hence, unlike the Multivariate Generalized Marshall–Olkin case, where independence is assumed among the original random variables $(E_S)_{S \in \mathscr{P}_0}$, here a dependence of Archimedean type is considered. Since here the generator is the Laplace transform of a positive random variable, we set in the case of Archimedean copulas with completely monotone generator (see McNeil and Nešlehová [20]).

The corresponding marginal survival distributions are

$$\bar{F}_{MGMO,\Lambda,j}(t) = L\left(-\ln \bar{F}_{MGMO,j}(t)\right) = \\ = L \circ K_j(t)$$

where

$$K_j(t) = \sum_{S: j \in S} H_S(t)$$

and the associated survival copula function is

$$\begin{aligned} \hat{C}_{MGMO,\Lambda}(\mathbf{u}) &= \hat{C}_{MGMO,\Lambda}\left(\mathbf{u}; (H_S)_{S \in \mathscr{P}_0}\right) = \\ &= L\left(-\ln \hat{C}_{MGMO}\left(e^{-L^{-1}(u_1)}, \ldots, e^{-L^{-1}(u_d)}\right)\right) = \\ &= L\left(\sum_{S \in \mathscr{P}_0} \max_{j \in S} H_S \circ K_j^{-1}(L^{-1}(u_j))\right) = \\ &= L\left(\sum_{S \in \mathscr{P}_0} H_S\left(\max_{j \in S} K_j^{-1} \circ L^{-1}(u_j)\right)\right) \end{aligned}$$

Remark 5.8 **The bivariate case** Let $d = 2$. If as in Example 5.4 $H_1 = H_{\{1\}}$, $H_2 = H_{\{2\}}$ and $H_3 = H_{\{1,2\}}$ we have that (5.23) takes the form

$$\bar{F}_{GMO,\Lambda}(x_1, x_2) = L(H_1(x_1) + H_2(x_2) + H_3(\max(x_1, x_2))) \tag{5.25}$$

and the associated survival copula function is

$$\hat{C}_{GMO,\Lambda}(u, v) = L\left(L^{-1}(u) + L^{-1}(v) - H_3\left(\min\left(K_1^{-1} \circ L^{-1}(u), K_2^{-1} \circ L^{-1}(v)\right)\right)\right) \tag{5.26}$$

where $K_i = H_i + H_3$, $i = 1, 2$.

Survival distribution functions of type (5.25) and copulas of type (5.26) represent a subset of the family of Archimedean-based Marshall–Olkin distributions and copulas, introduced in Mulinacci [22]: on the contrary to the case here considered, any Archimedean generator in (5.25) and (5.26) is there allowed. In that paper, some statistical properties (such as the Kendall's function and the Kendall's tau) of distributions and copulas of type (5.25) and (5.26) are studied.

An algorithm to build simulations of copulas of type (5.26) can be found in Durante [7].

Inheriting this property from the generalized Marshall–Olkin case, distributions of type (5.25) and copulas of type (5.26), are exchangeable if and only if $H_1 = H_2 = H$. In particular we get

$$\hat{C}_{GMO,\Lambda}(u, v; H, H, H_3) = L\left(L^{-1}(u) + L^{-1}(v) - H_3 \circ K^{-1} \circ L^{-1}(\max(u, v))\right)$$

where $K = H + H_3$.

Copulas of this type represent a particular specification of the copulas introduced in Durante et al. [8], defined as

$$C_{\phi,\psi}(u, v) = \phi^{[-1]}(\phi(u \wedge v) + \psi(u \vee v))$$

with $\phi : [0, 1] \to [0, +\infty]$, continuous, convex and strictly decreasing, $\psi : [0, 1] \to [0, +\infty]$, continuous, decreasing and such that $\psi(1) = 0$ and $\psi - \phi$ increasing in $[0, 1]$. The Mix-GMO exchangeable copula is recovered by setting $\phi(1) = L^{-1}(1) = 0$ and $\psi(t) = H \circ K^{-1} \circ L^{-1}(t)$.

Remark 5.9 The Mix-MGMO copulas family contains the SMMO family as a proper subset. In fact, we recover the copula functions of Example 5.3 considering, for $S \in \mathscr{P}_0$, $H_S = \lambda_S H$ for some H and $\lambda_S \geq 0$ with

$$\bar{\lambda}_j = \sum_{S:j \in S} \lambda_S > 0.$$

In fact, in this case,

$$\bar{F}_{MGMO,\Lambda}(\mathbf{t}) = L\left(\sum_{S \in \mathscr{P}_0} \lambda_S H\left(\max_{j \in S} t_j\right)\right),$$

and, since $K_j(x) = \bar{\lambda}_j H(t)$,

$$\bar{F}_{MGMO,\Lambda,j}(t) = L\left(\bar{\lambda}_j H(t)\right)$$

and

$$\hat{C}_{MGMO,\Lambda}(\mathbf{u}) = L\left(\sum_{S \in \mathscr{P}_0} \lambda_S \max_{j \in S} \frac{L^{-1}(u_j)}{\bar{\lambda}_j}\right)$$

Notice that this copula does not depend on H and coincides with the SMMO copula of Example 5.3.

A subclass of SMMO distributions and copula functions is considered in Cherubini and Mulinacci [5] where the model is applied to study the systemic risk and the contagion effects among the banks in a country.

The subclass of exchangeable SMMO copulas can be obtained alternatively as a frailty model as shown in Mai et al. [17]. Their approach works as follows: let $p(t)$ be a given distribution function on $[0,+\infty)$ $\epsilon_1, \epsilon_2, \ldots, \epsilon_d$ be i.i.d. unit exponentially distributed random variables, $Y > 0$ be a random variable with Laplace transform $L(x)$ and $\Lambda_t \neq 0$ be a Lévy subordinator with Laplace exponent Ψ; assuming that all the involved random variables and the Lévy subordinator are mutually independent, set

$$Y_t = \Lambda_{YL^{-1}(1-p(t))/\Psi(1)}$$

and define

$$\tau_k = \inf\{t \geq 0 : Y_t \geq \epsilon_k\}.$$

The survival copula associated to $(\tau_1, \tau_2, \ldots, \tau_d)$ is

$$C(\mathbf{u}) = L\left(\sum_{i=1}^{d} L^{-1}(u_{(i)})\frac{\Psi(i) - \Psi(i-1)}{\Psi(1)}\right)$$

where $u_{(1)} \leq \cdots \leq u_{(d)}$ denotes the ordered list of $u_1, \ldots, u_d$.

Since according to Remark 5.4 the power-mixing technique preserves the lower orthant order, we have that from (5.19)

$$\hat{C}_{MGMO,\Lambda}(\mathbf{u}) \geq L\left(\sum_{i=1}^{d} L^{-1}(u_i)\right).$$

Moreover, if, given two sets of generating functions $(H_S)_{S\in\mathscr{P}_0}$ and $(V_S)_{S\in\mathscr{P}_0}$, one of the conditions **C1**, **C2** or **C3** of Sect. 5.4 is satisfied then

$$\hat{C}_{MGMO,\Lambda}\left(\mathbf{u}, (H_S)_{S\in\mathscr{P}_0}\right) \geq \hat{C}_{MGMO,\Lambda}\left(\mathbf{u}, (V_S)_{S\in\mathscr{P}_0}\right)$$

According to the results presented in Sect. 5.2.3, the residual life of a vector $\tau = (\tau_1, \ldots, \tau_d)$, whose survival joint distribution is Mix-MGMO,

$$\tau_s = [\tau_1 - s, \ldots, \tau_d - s | \tau_1 > s, \ldots, \tau_d > s] \text{ for } s > 0$$

has a distribution that is again a mixture. More precisely, by Proposition 5.1 and (5.21), assuming $\mathbf{s} = (s, \ldots, s)$

$$\bar{F}_{\tau_s}(\mathbf{t}) = L_{\mathbf{s}}\left(\sum_{S\in\mathscr{P}_0} W_{S,s}\left(\max_{j\in S} t_j\right)\right)$$

and by (5.22)

$$\hat{C}_{\tau_s}(\mathbf{u}) = L_{\mathbf{s}}\left(\sum_{S\in\mathscr{P}_0}\max_{j\in S} W_{S,s}\circ\left(\sum_{J:j\in J} W_{J,s}\right)^{-1}(L_{\mathbf{s}}^{-1}(u_j))\right).$$

Example 5.6 If Λ is $\frac{1}{\theta}$-stable with $\theta \geq 1$, its Laplace transform is $L(x) = e^{-x^{1/\theta}}$ which is the generator of the Gumbel copula. In this case,

$$\bar{F}_{MGMO,\theta}(\mathbf{t}) = \exp\left(-\left(\sum_{S\in\mathscr{P}_0} H_S\left(\max_{j\in S} t_j\right)\right)^{1/\theta}\right)$$

and

$$\hat{C}_{MGMO,\theta}(\mathbf{u}) = \exp\left(-\left(\sum_{S\in\mathscr{P}_0} H_S\left(\max_{j\in S} K_j^{-1}((-\ln u_j)^{\theta})\right)\right)^{1/\theta}\right).$$

If, in particular, $H_S = \lambda_S H$ for some H and $\lambda_S \geq 0$ with

$$\bar{\lambda}_S = \sum_{S:j\in S}\lambda_S > 0,$$

then

$$\bar{F}_{MGMO,\theta}(\mathbf{t}) = \exp\left(\left(\sum_{S\in\mathscr{P}_0}\lambda_S H\left(\max_{j\in S} t_j\right)\right)^{1/\theta}\right),$$

$$\bar{F}_{MGMO,\theta,j}(t) = \exp\left(-\bar{\lambda}_j^{1/\theta} H^{1/\theta}(t)\right)$$

and

$$\hat{C}_{MGMO,\theta}(\mathbf{u}) = \exp\left(-\left(\sum_{S\in\mathscr{P}_0}\lambda_S\max_{j\in S}\frac{(-\ln u_j)^{\theta}}{\bar{\lambda}_j}\right)^{1/\theta}\right).$$

References

1. Bernhart, G., Mai, J.F., Scherer, M.: On the construction of low-parametric families of min-stable multivariate exponential distributions in large dimensions. Depend. Model. **3**, 29–46 (2015)
2. Bernhart, G., Escobar, M., Mai, J.F., Scherer, M.: Default models based on scale mixtures of Marshall-Olkin copulas: properties and applications. Metrika **76**(2), 179–203 (2013)
3. Charpentier, A.: Dynamic dependence ordering for Archimedean copulas and distorted copulas. Kybernetika, **44**(6), 777–794 (2008)
4. Charpentier, A., Fougères, A.-L., Genest, C., Nešlehová, J.G.: Multivariate Archimax copulas. J. Multivar. Anal. **126**, 118–136 (2014)
5. Cherubini, U., Mulinacci, S.: Systemic risk with exchangeable contagion: application to the European banking systems. Working paper: http://arxiv.org/abs/1502.01918 (2014)
6. Durante, F., Girard, S., Mazo, G.: Copulas based on Marshall-Olkin machinery. In: Cherubini, U., Durante, F., Mulinacci, S. (eds.) Marshall-Olkin Distributions—Advances in Theory and Applications. Springer Proceedings in Mathematics and Statistics. Springer International Publishing Switzerland (2015)
7. Durante, F.: Simulating from a family of generalized Archimedean copulas. In: Melas, V.B., Mignani, S., Monari, P., Salmaso, L. (eds.) Topics in Statistical Simulation. Springer Proceedings in Mathematics, pp. 149–156. Springer, New York (2014)
8. Durante, F., Quesada-Molina, J.J., Sempi, C.: A generalization of the Archimedean class of bivariate copulas. Ann. Inst. Stat. Math. **59**(3), 487–498 (2007)
9. Frostig, E., Pellerey, F.: General Marshall-Olkin models, dependence orders and comparisons of environmental processes. In: Cherubini, U., Durante, F., Mulinacci, S. (eds.) Marshall-Olkin Distributions—Advances in Theory and Applications. Springer Proceedings in Mathematics and Statistics. Springer International Publishing Switzerland (2015)
10. Joe, H.: Families of min-stable multivariate exponential and multivariate extreme value distributions. Stat. Probab. Lett. **9**(1), 75–81 (1990)
11. Joe, H.: Multivariate Models and Multivariate Dependence Concepts. Chapman & Hall, London (1997)
12. Joe, H., Hu, T.: Multivariate distributions from mixtures of max-infinitely divisible distributions. J. Multivar. Anal. **57**, 240–265 (1996)
13. Li, H.: Tail dependence comparison of survival Marshall-Olkin copulas. Methodol. Comput. Appl. Probab. **10**(1), 39–54 (2008)
14. Li, H.: Orthant tail dependence of multivariate extreme value distributions. J. Multivar. Anal. **100**(1), 243–256 (2009)
15. Lin, J., Li, X.: Multivariate generalized Marshall-Olkin distributions and copulas. Methodol. Comput. Appl. Probab. **16**(1), 53–78 (2014)
16. Li, X., Pellerey, F.: Generalized Marshall-Olkin distributions and related bivariate aging properties. J. Multivar. Anal. **102**, 1399–1409 (2011)
17. Mai, J.F., Scherer, M., Zagst, R.: CIID frailty models and implied copulas. Copulae in Mathematical and Quantitative Finance. Lecture Notes in Statistics 2013, pp. 201–230. Springer, New York (2013)
18. Marshall, A.W., Olkin, I.: A multivariate exponential distribution. J. Am. Stat. Assoc. **62**, 30–49 (1967)
19. Marshall, A.W., Olkin, I.: Families of multivariate distributions. J. Am. Stat. Assoc. **83**, 834–841 (1988)
20. McNeil, A.J., Nešlehová, J.G.: Multivariate Archimedean copulas, d-monotone functions and L1-norm symmetric distributions. Ann. Stat. **37**, 3059–3097 (2009)
21. Muliere, P., Scarsini, M.: Characterization of a Marshall-Olkin type class of distributions. Ann. Inst. Stat. Math. **39**, 429–441 (1987)
22. Mulinacci, S.: Archimedean-based Marshall-Olkin distributions and related copula functions. Working paper: http://arxiv.org/abs/1502.01912 (2014)

Chapter 6
Extended Marshall–Olkin Model and Its Dual Version

Jayme Pinto and Nikolai Kolev

Abstract We propose an extension of the generalized bivariate Marshall–Olkin model assuming dependence between the random variables involved. Probabilistic, aging properties, and survival copula representation of the extended model are obtained and illustrated by examples. Bayesian analysis is performed and possible applications are discussed. A dual version of extended Marshall–Olkin model is introduced and related stochastic order comparisons are presented.

6.1 Introduction

A variety of bivariate (multivariate) extensions of the univariate exponential distribution have been considered in the literature. These include the distributions of [6, 10, 11, 24], see a full review in [2].

The vector (X_1, X_2) meets the Marshall–Olkin model (MO hereafter) whenever it admits the stochastic representation

$$(X_1, X_2) = [\min(T_1, T_3), \min(T_2, T_3)], \tag{6.1}$$

where the random variables T_i are independent and exponentially distributed with parameters $\lambda_i > 0$, respectively, $i = 1, 2, 3$.

The random variables T_1 and T_2 in (6.1) can be interpreted as the arrival time of individual shocks for two different components, while T_3 represents the time of arrival of a shock common to both components. Several applications of (6.1) can be mentioned. In reliability theory it can describe the lifetime of a system of two components operating in a random environment and subject to three independent

J. Pinto (✉) · N. Kolev
Department of Statistics, University of São Paulo, Sao Paulo, Brazil
e-mail: jaymeaugusto@gmail.com

N. Kolev
e-mail: kolev.ime@gmail.com

U. Cherubini et al. (eds.), *Marshall–Olkin Distributions - Advances in Theory and Applications*, Springer Proceedings in Mathematics & Statistics 141,
DOI 10.1007/978-3-319-19039-6_6

sources of damage; in actuarial science (6.1) can model the survival function of a married couple in a contract of joint life with last-survivor annuity, see [39].

Let us consider a vector (X_1, X_2) of nonnegative continuous random variables with joint survival function $S_{X_1,X_2}(x_1, x_2) = P(X_1 > x_1, X_2 > x_2)$. The bivariate survival function of (X_1, X_2) corresponding to (6.1) is

$$S_{X_1,X_2}(x_1, x_2) = \exp\{-\lambda_1 x_1 - \lambda_2 x_2 - \lambda_3 \max(x_1, x_2)\}, \quad x_1, x_2 \geq 0 \tag{6.2}$$

and defines the MO bivariate exponential distribution, see [24].

Among the different bivariate lifetime models, the MO bivariate exponential distribution (6.2) is the most popular one. Since $P(X_1 = X_2) = \frac{\lambda_3}{\lambda_1+\lambda_2+\lambda_3} > 0$, it contains a singular component and has been used to model datasets with ties. In addition, (6.2) possesses the bivariate lack-of-memory property. The MO bivariate exponential distribution is widely used in risk management, see [25] and Chap. 3 in [23].

The MO distribution (6.2) is the only bivariate exponential distribution with exponential marginals, i.e., having decreasing density and constant hazard functions. If one of the marginal empirical density is not decreasing, one could not use (6.2) as a correct model. For this reason, in [24] Weibull distribution is suggested for T_1 and T_2. Many generalizations of MO-type distributions have been proposed. For example, consult [30] and several very recent contributions by [17, 18, 27, 28, 37] and references therein. Usually the authors assume various kinds of distribution for T_1, T_2, and T_3 in (6.1), but always keep them independent.

To overcome the restriction of exponential marginals assumed in [24], it was proposed in [21] the Generalized Marshall–Olkin distribution (to be abbreviated GMO) assuming only independence among the nonnegative continuous random variables T_1, T_2, and T_3 in (6.1).

Let the lifetimes T_i have survival function $S_{T_i}(x_i) = P(T_i > x_i)$, density $f_{T_i}(x_i)$, and failure (hazard) rate $r_{T_i}(x_i) = \frac{f_{T_i}(x_i)}{S_{T_i}(x_i)}$. Denote by $H_{T_i}(x) = \int_0^x r_{T_i}(t)dt$ their right continuous cumulative failure rate functions satisfying

$$H_{T_i}(0) = 0, \; H_{T_i}(\infty) = \infty \quad \text{and} \quad H_{T_i}(x_i) < \infty \quad \text{for} \quad x_i > 0, \; i = 1, 2, 3.$$

In [21] the GMO model is defined by

$$\begin{aligned} S_{X_1,X_2}(x_1, x_2) &= S_{T_1}(x_1) S_{T_2}(x_2) S_{T_3}(\max(x_1, x_2)) \\ &= \exp\{-H_{T_1}(x_1) - H_{T_2}(x_2) - H_{T_3}(\max(x_1, x_2))\}. \end{aligned} \tag{6.3}$$

The second representation is a consequence of the well-known relation $S_{T_i}(x) = \exp\{-H_{T_i}(x)\}$, see [3]. Despite bringing more flexibility and enlarging the range of applications of MO-type models, the independence assumption in GMO model may not be satisfied in the following situations:

1. Two electric machines operating in the same factory and sharing the same maintenance policy, both subject to blackouts;

2. Two components in the structure of the same building sharing reverse load or reverse set of stresses, both subject to earthquakes.

One can easily identify dependence between the individual "shocks" in the preceding examples: it is positive in case 1 and negative in case 2. Taking into account this real possibility, our aim is to define a MO-type model that incorporates dependence between the random variables T_1 and T_2 while preserving the stochastic representation (6.1).

In Sect. 6.2 we describe our proposal for a possible extension of the MO and GMO models (6.1) and (6.3), and analyze related basic properties. We derive the copula expression for the extended MO model, provide examples, and analyze the distribution of the residual lifetimes.

Extensive work has been done in developing inference procedures for MO-type models, see [14, 16] in this respect. Similar approach can be applied for the extended MO model, but our choice is to perform a Bayesian data analysis in Sect. 6.3 to demonstrate its advantage. We introduce a dual version of the extended MO model in Sect. 6.4 based on max instead of min operation in (6.1) and outline its properties. Related stochastic order comparisons are presented as well. We finish with a discussion, including admissible-related research and applied directions.

6.2 Extended Marshall–Olkin Model

Let us consider a pair (T_1, T_2) of nonnegative continuous random variables with a joint survival function of the form

$$S_{T_1,T_2}(x_1, x_2) = P(T_1 > x_1, T_2 > x_2) \quad \text{having marginals} \quad S_{T_1}(x_1) \text{ and } S_{T_2}(x_2).$$

In addition, assume that the random variable T_3 with survival function $S_{T_3}(x) = P(T_3 > x)$ is independent of T_1 and T_2 in stochastic representation (6.1). Hence, we arrive to the following

Definition 6.1 The extended MO (EMO) model is defined by its survival function

$$\begin{aligned} S_{X_1,X_2}(x_1, x_2) &= P(T_1 > x_1, T_2 > x_2, T_3 > \max\{x_1, x_2\}) \\ &= S_{T_1,T_2}(x_1, x_2) S_{T_3}(\max\{x_1, x_2\}). \end{aligned} \tag{6.4}$$

Observe that GMO model (6.3) is a particular case of the EMO model (6.4) when T_1 and T_2 are independent. If in addition the random variables T_1, T_2, and T_3 are exponentially distributed with parameters λ_1, λ_2, and λ_3, respectively, one gets the classical MO bivariate exponential distribution (6.2). Figure 6.1 illustrates the relationship among the three models.

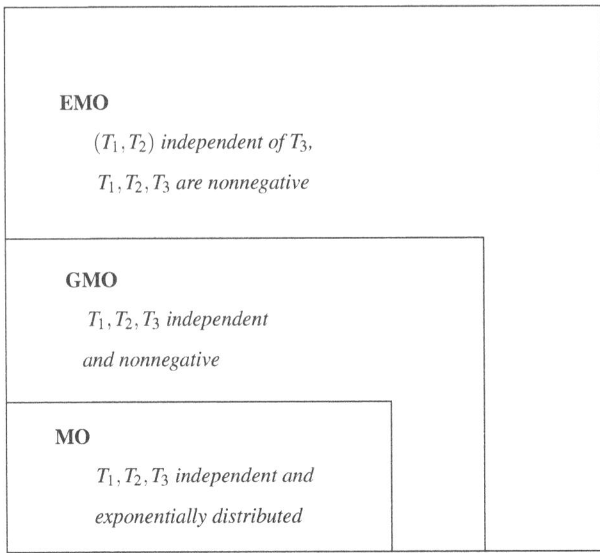

Fig. 6.1 MO, GMO, and EMO models compared

6.2.1 Alternative Representation of the EMO Model

In order to investigate the properties of the EMO model, the exponential representation of a bivariate survival function turns out to be useful. Recall that any bivariate survival function can be specified by

$$S_{T_1,T_2}(x_1,x_2) = \exp\{-H_{T_1}(x_1) - H_{T_2}(x_2) + H_{T_1,T_2}(x_1,x_2)\}, \qquad (6.5)$$

see Sect. 2.2 in [7]. In the last relation $H_{T_i}(x_i)$ is the cumulative hazard of T_i, $i = 1, 2$ and the function $H_{T_1,T_2}(x_1,x_2)$ is defined by

$$H_{T_1,T_2}(x_1,x_2) = \ln\left\{\frac{S_{T_1,T_2}(x_1,x_2)}{S_{T_1}(x_1)S_{T_2}(x_2)}\right\}, \qquad (6.6)$$

satisfying the boundary conditions $H_{T_1,T_2}(0,x_2) = H_{T_1,T_2}(x_1,0) = 0$ for all $x_1, x_2 \geq 0$. Equation (6.5) can be viewed as a bivariate version of the univariate exponential representation of survival function, e.g., $S_{T_i}(x) = \exp\{-H_{T_i}(x)\}$, $i = 1, 2$.

According to [38], any bivariate survival function can be decomposed as a product of marginal survival functions and a function $\Omega(x_1,x_2)$ via

$$S_{T_1,T_2}(x_1,x_2) = S_{T_1}(x_1)S_{T_2}(x_2)\Omega(x_1,x_2).$$

The multiplier $\Omega(x_1, x_2)$ is known as Sibuya's dependence function of the random vector (T_1, T_2) and obviously

$$\Omega(x_1, x_2) = \frac{S_{X_1,X_2}(x_1, x_2)}{S_{X_1}(x_1)S_{X_2}(x_2)} = \exp\{H_{T_1,T_2}(x_1, x_2)\}.$$

The last relation shows that $H_{T_1,T_2}(x_1, x_2)$ can be interpreted as the free of marginal influence contribution to the genuine dependence between T_1 and T_2.

Taking into account (6.4) and (6.5), one can equivalently define the EMO model in terms of cumulative failure rates $H_{T_i}, i = 1, 2, 3$ and the function H_{T_1,T_2} as

$$S_{X_1,X_2}(x_1, x_2) = \exp\{-H_{T_1}(x_1) - H_{T_2}(x_2) - H_{T_3}(\max(x_1, x_2)) + H_{T_1,T_2}(x_1, x_2)\}. \tag{6.7}$$

It is direct to check that the marginal survival functions of $S_{X_1,X_2}(x_1, x_2)$ are given by $S_{X_i}(x_i) = \exp\{-H_{T_i}(x_i) - H_{T_3}(x_i)\},\ i = 1, 2.$

Remark 6.1 Note that the only difference between GMO and EMO models, given in (6.3) and (6.7), is the presence of the function $H_{T_1,T_2}(x_1, x_2)$ in (6.7). In fact, the equation $H_{T_1,T_2}(x_1, x_2) = 0$ (satisfied for all $x_1, x_2 \geq 0$) characterizes the independence between random variables T_1 and T_2.

6.2.2 *Probabilistic Properties*

We mention several probabilistic properties of the EMO model (6.7) formalized in Definition 6.1.

6.2.2.1 Positive and Negative Quadrant Dependence

Recall that a nonnegative random vector (X_1, X_2) is positive quadrant dependent (PQD) if $P(X_1 \leq x_1, X_2 \leq x_2) \geq P(X_1 \leq x_1)P(X_2 \leq x_2)$, or equivalently, one gets $P(X_1 > x_1, X_2 > x_2) \geq P(X_1 > x_1)P(X_2 > x_2)$, for all $x_1, x_2 \geq 0$. The vector (X_1, X_2) is negative quadrant dependent (NQD) when the last two relations are valid with the inequality sign reversed, see [19].

Remark 6.2 Note that X_1 and X_2 defined by stochastic representation (6.1) are associated random variables since they are increasing functions (e.g., min) of random variables T_1, T_2, and T_3. Hence, X_1 and X_2 are PQD in the MO and GMO models (6.2) and (6.3), according to Theorem 2.2 and Property P_3 given on page 30 in [3]. Thus, the statement of Proposition 2.1 in [21] follows without the need of obtaining the copula corresponding to the GMO distribution (6.3). The same conclusion cannot be handled for the EMO model since the vector (T_1, T_2) may be NQD and not associated, therefore.

The next statement characterizes the NQD property of the EMO model.

Theorem 6.1 *The vector* (X_1, X_2) *following the EMO model is NQD if and only if*

$$S_{T_1,T_2}(x_1, x_2) \leq S_{T_1}(x_1)S_{T_2}(x_2)S_{T_3}(\min\{x_1, x_2\}), \tag{6.8}$$

or equivalently,

$$H_{T_1,T_2}(x_1, x_2) + H_{T_3}(\min(x_1, x_2)) \leq 0$$

for all $x_1, x_2 \geq 0$.

Proof The vector (X_1, X_2) is NQD if and only if $S_{X_1,X_2}(x_1, x_2) \leq S_{X_1}(x_1)S_{X_2}(x_2)$, for all $x_1, x_2 \geq 0$. The necessary and sufficient condition given by relation (6.8) can be obtained using (6.7), relations $S_{X_i}(x_i) = \exp\{-H_{T_i}(x_i) - H_{T_3}(x_i)\}$, $i = 1, 2$, and the fact that $\min(x_1, x_2) = x_1 + x_2 - \max(x_1, x_2)$.

The second inequality is a direct consequence of (6.8). □

Remark 6.3 Note that the EMO model (6.7) may be not PQD or not NQD, for all $x_1, x_2 \geq 0$, conditional on the distributional parameters involved. Such a case (with a "local" PQD and "local" NQD property) is illustrated in Example 6.2.

6.2.2.2 Symmetry, Asymmetry, and Bounds for the Joint Survival Function

The MO model (6.2) may be exchangeable or not, depending on the parameters λ_1 and λ_2. If T_1 and T_2 are exponentially distributed with the same parameter then the MO model is always symmetric. The same happens with the GMO model (6.3): if T_1 and T_2 are identically distributed then the GMO model is exchangeable. Both cases are analogous because the copula that joins T_1 and T_2 is the independence one, which is symmetric.

Interestingly, if one starts with identically distributed random variables T_1 and T_2 in the EMO model (6.7), but (T_1, T_2) are connected by a nonexchangeable copula, then the resulting EMO model will be nonexchangeable. Thus, the EMO model is not necessarily exchangeable as demonstrated in the next example.

Example 6.1 (Nonexchangeable EMO model with identically distributed marginals) Consider the stochastic representation (6.1) and let T_i be exponentially distributed with parameter $\lambda_i = 1$, i.e., $H_{T_i}(x_i) = x_i$, $i = 1, 2, 3$. Then X_1 and X_2 are exponentially distributed with a common parameter $\lambda = 2$, i.e., $H_{X_i}(x_i) = 2x_i$, $i = 1, 2$. Let the asymmetric copula that joins (T_1, T_2) be given by

$$C_{T_1,T_2}(u, v) = uv + uv(1-u)(1-v)[(a-b)v(1-u) + b],$$

where $|b| \leq 1$, $\frac{b-3-\sqrt{9+6b-3b^2}}{2} \leq a \leq 1$ and $a \neq b$, see Example 3.16 in [31]. As a result, the corresponding EMO joint survival function is given by

$$S_{X_1,X_2}(x_1, x_2) = \left[\exp\{-x_1 - x_2 - \max(x_1, x_2)\}\right]$$

$$\times\left[1 + (1 - \exp\{-x_1\})(1 - \exp\{-x_2\})[(a - b)(1 - \exp\{-x_2\})\exp\{-x_1\} + b]\right],$$

being asymmetric.

Besides the knowledge of the distributions of T_1, T_2, and T_3, a key aspect for deriving EMO models is the knowledge of the joint survival function of T_1 and T_2 given by (6.5). An important component in (6.5) is the function $H_{T_1,T_2}(x_1, x_2)$ defined by (6.6), which may serve as a measure of dependence between random variables T_1 and T_2.

In the case of incomplete information, it is still possible to obtain bounds for the survival function of the EMO model (6.7) based on the knowledge of the marginal survival functions S_{T_i} or their cumulative failure rate functions H_{T_i}, $i = 1, 2, 3$. The following statement holds.

Lemma 6.1 *The lower and upper bounds for the survival function of the EMO model (6.7), are given by*

$$L(x_1, x_2) \leq S_{X_1,X_2}(x_1, x_2) \leq U(x_1, x_2),$$

where

$$L(x_1, x_2) = \max\left\{\left[\exp\{-H_{T_1}(x_1)\} + \exp\{-H_{T_2}(x_2)\} - 1\right], 0\right\}\exp\{-H_{T_3}(\max(x_1, x_2))\}$$

and

$$U(x_1, x_2) = \min\left\{\exp\{-H_{T_1}(x_1)\}, \exp\{-H_{T_2}(x_2)\}\right\}\exp\{-H_{T_3}(\max(x_1, x_2))\}.$$

Remark 6.4 Since $L(x_1, x_2) > \max(S_{X_1}(x_1) + S_{X_2}(x_2) - 1, 0)$, as well as $U(x_1, x_2) < \min(S_{X_1}(x_1), S_{X_2}(x_2))$, the bounds obtained in Lemma 6.1 are *sharper* than the usual Fréchet–Hoeffding bounds.

6.2.3 Survival Copula of the EMO Model

By Sklar's theorem, the dependence structure of a random vector (X_1, X_2) with joint survival function $S_{X_1,X_2}(x_1, x_2)$ and continuous marginal survival functions $S_{X_1}(x_1)$ and $S_{X_2}(x_2)$ has unique survival copula

$$\overline{C}_{X_1,X_2}(u, v) = S_{X_1,X_2}(S_{X_1}^{-1}(u), S_{X_2}^{-1}(v)), \quad (u, v) \in [0, 1],$$

where $S_{X_i}^{-1}$ is the right continuous inverse of S_{X_i}, $i = 1, 2$, see [31]. The triplet $(S_{X_1}, S_{X_2}, \overline{C}_{X_1,X_2})$ allows to analyze the dependence properties between X_1 and X_2.

Recall that the marginal survival functions of $S_{X_1,X_2}(x_1, x_2)$ from (6.7) are given by $S_{X_i}(x_i) = \exp\{-H_{T_i}(x_i) - H_{T_3}(x_i)\}$. The following statement holds.

Lemma 6.2 *Set* $G_i(x_i) = H_{T_i}(x_i) + H_{T_3}(x_i)$, $i = 1, 2$. *The survival copula* $\overline{C}_{X_1,X_2}(u, v)$ *of the EMO model is given by*

$$\overline{C}_{X_1,X_2}(u, v) = \begin{cases} uv\exp\{H_{T_3}(G_2^{-1}(-\ln v)) + G(u, v)\}, & \text{if } G_1^{-1}(-\ln u) > G_2^{-1}(-\ln v); \\ uv\exp\{H_{T_3}(G_1^{-1}(-\ln u)) + G(u, v)\}, & \text{if } G_1^{-1}(-\ln u) \le G_2^{-1}(-\ln v), \end{cases} \tag{6.9}$$

where $u, v \in (0, 1)$ *and* $G(u, v) = H_{T_1,T_2}(S_{X_1}^{-1}(u), S_{X_2}^{-1}(v))$.

Proof First observe that $S_{X_1}(x_1) = \exp\{-H_{T_1}(x_1) - H_{T_3}(x_1)\} = \exp\{-G_1(x_1)\}$. Solving $S_{X_1}(x_1) = u$ we get $S_{X_1}^{-1}(u) = x_1$. By analogy, from $\exp\{-G_1(x_1)\} = u$ we obtain $x_1 = G_1^{-1}(-\ln u)$ and therefore

$$S_{X_1}^{-1}(u) = G_1^{-1}(-\ln u), \quad \text{i.e.,} \quad G_1(S_{X_1}^{-1}(u)) = -\ln u.$$

In a similar way we obtain

$$S_{X_2}^{-1}(v) = G_2^{-1}(-\ln v), \quad \text{i.e.,} \quad G_2(S_{X_2}^{-1}(v)) = -\ln v. \tag{6.10}$$

If $x_1 > x_2 \ge 0$, i.e., $S_{X_1}^{-1}(u) > S_{X_2}^{-1}(v)$, then $G_1^{-1}(-\ln u) > G_2^{-1}(-\ln v)$.

Let $\overline{C}_{X_1,X_2}(u, v)$ be the survival copula corresponding to $S_{X_1,X_2}(x_1, x_2)$, i.e.,

$$\ln[\overline{C}_{X_1,X_2}(u, v)] = \ln[S_{X_1,X_2}(S_{X_1}^{-1}(u), S_{X_2}^{-1}(v))], \quad u, v \in (0, 1).$$

Therefore, using relation (6.7) we have

$$\ln[\overline{C}_{X_1,X_2}(u, v)] = \ln[\exp\{-H_{T_1}(S_{X_1}^{-1}(u)) - H_{T_2}(S_{X_2}^{-1}(v)) - H_{T_3}(S_{X_1}^{-1}(u)) + G(u, v)\}],$$

with $G(u, v) = H_{T_1,T_2}(S_{X_1}^{-1}(u), S_{X_2}^{-1}(v))$, which is equivalent to

$$\ln[\overline{C}_{X_1,X_2}(u, v)] = \ln u - H_{T_2}(S_{X_2}^{-1}(v)) + G(u, v).$$

Due to (6.10) we get

$$\ln \overline{C}_{X_1,X_2}(u, v) = \ln(uv) + H_{T_3}(S_{X_2}^{-1}(v)) + G(u, v).$$

Finally, $\overline{C}_{X_1,X_2}(u, v) = uv\exp\{H_{T_3}(G_2^{-1}(-\ln v)) + G(u, v)\}$ if $S_{X_1}^{-1}(u) > S_{X_2}^{-1}(v)$.

By analogy, when $0 \le x_1 \le x_2$, i.e., $G_1^{-1}(-\ln u) \le G_2^{-1}(-\ln v)$, one obtains

$$\overline{C}_{X_1,X_2}(u,v) = uv \exp\{H_{T_3}(G_2^{-1}(-\ln v)) + G(u,v)\}$$

and representation (6.9) is established. □

Remark 6.5 The function $\exp\{G(u,v)\}$ is the only product extra multiplier in (6.9) in addition to the copula expression corresponding to GMO distribution (compare with Remark 6.1) and (2.3) by [21]. This extra term permits "local" NQD modeling of (X_1, X_2), as shown in Example 6.2.

The following example illustrates how the survival copula given in (6.9) can be obtained. We offer quadrant dependence analysis as well.

Example 6.2 (*Survival copula of the EMO model and quadrant dependence analysis*) Assume that (T_1, T_2) is Gumbel's type I bivariate exponentially distributed with unit exponential marginals, see [11]. The corresponding survival function is given by

$$S_{T_1,T_2}(x_1,x_2) = \exp(-x_1 - x_2 - \theta x_1 x_2),$$

where $x_1, x_2 \ge 0$ and $\theta \in [0,1]$. Let T_3 be independent of (T_1, T_2) with survival function $S_{T_3}(x) = \exp(-\lambda x)$, $\lambda > 0$. Following the above notations, we have $H_{T_1}(x) = H_{T_2}(x) = x$, $H_{T_3}(x) = \lambda x$ and $H_{T_1,T_2}(x_1,x_2) = -\theta x_1 x_2$. Therefore,

$$G_1(x_1) = H_{T_1}(x_1) + H_{T_3}(x_1) = (1+\lambda)x_1 \text{ and } G_2(x_2) = H_{T_2}(x_2) + H_{T_3}(x_2) = (1+\lambda)x_2.$$

The inverse functions are given by

$$G_1^{-1}(u) = \frac{u}{1+\lambda} \quad \text{and} \quad G_2^{-1}(v) = \frac{v}{1+\lambda}.$$

Since $\ln(x)$ is an increasing function and $1+\lambda > 0$, we have the following set of equivalent inequalities when $0 < u < v \le 1$:

$$G_1^{-1}(-\ln(u)) > G_2^{-1}(-\ln(v)) \Leftrightarrow \frac{-\ln(u)}{1+\lambda} > \frac{-\ln(v)}{1+\lambda} \Leftrightarrow \frac{\ln(u)}{1+\lambda} < \frac{\ln(v)}{1+\lambda}.$$

But $G(u,v) = H_{T_1,T_2}\left(G_1^{-1}(-\ln(u)), G_2^{-1}(-\ln(v))\right) = -\theta \frac{\ln(u)\ln(v)}{(1+\lambda)^2}$ and we obtain the survival copula of the corresponding EMO model

$$\overline{C}_{X_1,X_2}(u,v) = \begin{cases} uv \exp\left\{-\lambda \frac{\ln(v)}{1+\lambda} - \theta \frac{\ln(u)\ln(v)}{(1+\lambda)^2}\right\}, & \text{if } 0 < u < v \le 1; \\ uv \exp\left\{-\lambda \frac{\ln(u)}{1+\lambda} - \theta \frac{\ln(u)\ln(v)}{(1+\lambda)^2}\right\}, & \text{if } 1 \ge u \ge v > 0. \end{cases}$$

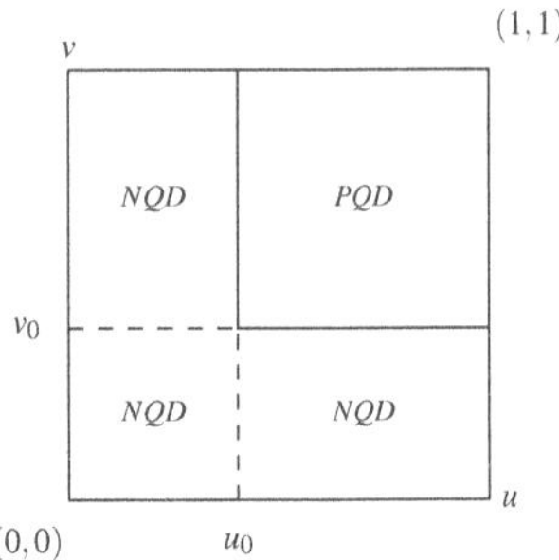

Fig. 6.2 Dependence analysis of the EMO model in Example 6.2

When $0 < u < v \leq 1$, we have $\overline{C}_{X_1,X_2}(u,v) \leq uv$ if

$$\exp\left\{-\lambda\frac{\ln(v)}{1+\lambda} - \theta\frac{\ln(u)\ln(v)}{(1+\lambda)^2}\right\} \leq 1, \quad \text{i.e.,} \quad 0 < u \leq u_0 = \exp\left\{-\frac{\lambda(1+\lambda)}{\theta}\right\}.$$

By analogy, when $0 < v \leq u \leq 1$ and $\overline{C}_{X_1,X_2}(u,v) \leq uv$ we get the inequality $0 < v \leq v_0 = \exp\left\{-\frac{\lambda(1+\lambda)}{\theta}\right\}$.

Therefore, as illustrated in Fig. 6.2, when $(u,v) \in \left[\exp\{-\frac{\lambda(1+\lambda)}{\theta}\}, 1\right]^2$ we have the "local" PQD property. Outside this set in the unit square, the "local" NQD property is valid.

6.2.4 Distributional Property for Residual Lifetimes

Many authors have studied the monotonicity in x_1 and x_2 of

$$P(X_1 > x_1 + t, X_2 > x_2 + t \mid X_1 > t, X_2 > t) = \frac{S_{X_1,X_2}(x_1+t, x_2+t)}{S_{X_1,X_2}(t,t)}$$

with respect to $t \geq 0$. Under the bivariate lack-of-memory property the last relation only depends on x_1 and x_2. In general, this conditional probability is a nonincreasing function and represents the joint survival function of the residual lifetime vector

$$\mathbf{X}_t = (X_{1t}, X_{2t}) = [(X_1 - t, X_2 - t) \mid X_1 > t, X_2 > t] \tag{6.11}$$

corresponding to (X_1, X_2), where X_{1t} and X_{2t} are the marginal residual lifetimes at time $t \geq 0$. The problem has important applications in industry, medicine, finance, economics, insurance, see [25, 39].

The aging performance of the residual lifetime vector $\mathbf{X}_t$ is well studied when the dependence structure between X_1 and X_2 is described by the family of Archimedean copulas, for details see [5, 29]. Thus, the results obtained are valid for exchangeable random vectors, i.e., for the class of bivariate distributions with same marginal distributions, which is a limitation for practical needs and applications.

The GMO type distributions given by (6.3) do not possess the bivariate lack-of-memory property and for this reason [21] investigated the aging behavior and dependence properties of $\mathbf{X}_t$. In particular, the authors show that if (X_1, X_2) has a GMO distribution, then so does $\mathbf{X}_t$.

We are now interested to examine the properties of the residual lifetime vector $\mathbf{X}_t$ corresponding to (X_1, X_2) satisfying (6.7), i.e., under the EMO model. The result is given in the following theorem.

Theorem 6.2 *If* (X_1, X_2) *follows an EMO model, then so does its residual lifetime vector* $\mathbf{X}_t$ *for any* $t \geq 0$.

Proof The survival function of the residual lifetime vector (6.11) is

$$P(\mathbf{X}_t > (x_1, x_2)) = P(X_1 - t > x_1, X_2 - t > x_2 \mid X_1 > t, X_2 > t)$$

$$= P(\min\{T_1 - t, T_3 - t\} > x_1, \min\{T_2 - t, T_3 - t\} > x_2 \mid \min\{T_1, T_3\} > t, \min\{T_2, T_3\} > t).$$

Taking into account that in the EMO model (6.7) the random variable T_3 is independent of the vector (T_1, T_2) we get

$$P(\mathbf{X}_t > (x_1, x_2)) = \frac{P(T_1 > t + x_1, T_2 > t + x_2)P(T_3 > t + \max\{x_1, x_2\})}{P(T_1 > t, T_2 > t)P(T_3 > t)},$$

i.e.,

$$P(\mathbf{X}_t > (x_1, x_2)) = \frac{S_{T_1,T_2}(t + x_1, t + x_2)}{S_{T_1,T_2}(t, t)} \times \frac{S_{T_3}(t + \max\{x_1, x_2\})}{S_{T_3}(t)}.$$

The last relation means that

$$P(\mathbf{X}_t > (x_1, x_2)) = P(T_{1t} > x_1, T_{2t} > x_2)P(T_{3t} > \max\{x_1, x_2\}),$$

where $T_{it} = [T_i - t \mid T_1 > t, T_2 > t]$, $i = 1, 2$, and $T_{3t} = [T_3 - t \mid T_3 > t]$. Thus, for any $t \geq 0$ the residual lifetime vector $\mathbf{X}_t$ in the EMO model has a stochastic representation

$$\mathbf{X}_t = (X_{1t}, X_{2t}) = [\min\{T_{1t}, T_{3t}\}, \min\{T_{2t}, T_{3t}\}],$$

e.g., being in the form (6.1). This means that if the random vector (X_1, X_2) follows an EMO model with survival function given by (6.7) then the corresponding residual lifetime vector $\mathbf{X}_t$ also follows an EMO model. □

As a consequence of the Theorem 6.B.16(b) in [36], we conclude that (X_1, X_2) and $\mathbf{X}_t = (X_{1t}, X_{2t})$ have the same type of copula (i.e., given by relation (6.9)) even if they are generated by different triples (T_1, T_2, T_3) and $(T_{1t}, T_{2t}, T_{3t}), t \geq 0$ of random variables satisfying (6.1).

Remark 6.6 If T_3 is exponentially distributed and (T_1, T_2) possesses the bivariate lack-of-memory property, $S_{T_1,T_2}(x_1 + t, x_2 + t) = S_{T_1,T_2}(x_1, x_2)S_{T_1,T_2}(t, t)$, for all $x_1, x_2, t > 0$, (e.g., the bivariate exponential distribution in [6]), then (X_1, X_2) following (6.1) will exhibit the bivariate lack-of-memory property as well. Therefore, $\mathbf{X}_t$ and (X_1, X_2) will have the same distribution and copula, which are independent of t, being time invariant.

6.3 Bayesian Data Analysis with EMO Models

As we noted, the EMO-type bivariate distributions exhibit a singular component along the line $x_1 = x_2$. Therefore, EMO model can serve as a good candidate to fit datasets with ties. In general, such a fit would be better than the corresponding results obtained if one uses some absolutely continuous model. We perform here a Bayesian analysis applying *OpenBugs* software (free available on `www.openbugs.net/w/Downloads`) for a soccer data with ties analyzed by [26].

If the bivariate random vector (X_1, X_2) satisfies the EMO model (6.4), its survival function has no discrete component and admits the Lebesgue decomposition

$$S_{X_1,X_2}(x_1, x_2) = (1 - \alpha)S^{ac}_{X_1,X_2}(x_1, x_2) + \alpha S^{si}_{X_1,X_2}(\max\{x_1, x_2\}),$$

where $S^{ac}_{X_1,X_2}(x_1, x_2)$ is an absolutely continuous survival function, $S^{si}_{X_1,X_2}(\max\{x_1, x_2\})$ is the singular component with support on the set $\Gamma = \{(x_1, x_2) \in R^2 \,|x_1 = x_2 = x\}$ and $\alpha = P(X_1 = X_2) \in [0, 1]$.

If $\alpha = 0$ then the joint distribution of the random vector (X_1, X_2) is absolutely continuous. Let

$$(1 - \alpha) f^{ac}_{X_1,X_2}(x_1, x_2) = \frac{\partial^2}{\partial x_1 \partial x_2} S_{X_1,X_2}(x_1, x_2).$$

The value of α can be obtained integrating both sides of last relation to get

$$1 - \alpha = \int_0^{\infty} \int_0^{\infty} \frac{\partial^2}{\partial x_1 \partial x_2} S_{X_1,X_2}(x_1, x_2) \mathrm{d}x_1 \mathrm{d}x_2.$$

By its turn, the survival function of EMO models can be decomposed in terms of the survival functions of (T_1, T_2) and T_3 as

$$S_{X_1,X_2}(x_1,x_2)=\begin{cases} S_{T_1,T_2}(x_1,x_2)S_{T_3}(x_1), & \text{if } x_1\ge x_2\ge 0,\\ S_{T_1,T_2}(x_1,x_2)S_{T_3}(x_2), & \text{if } 0\le x_1<x_2,\end{cases}$$

and

$$(1-\alpha)f^{ac}_{X_1,X_2}(x_1,x_2)$$
$$=\begin{cases} S_{T_3}(x_1)\frac{\partial^2}{\partial x_1\partial x_2}S_{T_1,T_2}(x_1,x_2)+\frac{\partial}{\partial x_2}S_{T_1,T_2}(x_1,x_2)\frac{d}{dx_1}S_{T_3}(x_1), & \text{if } x_1>x_2\ge 0,\\ S_{T_3}(x_2)\frac{\partial^2}{\partial x_1\partial x_2}S_{T_1,T_2}(x_1,x_2)+\frac{\partial}{\partial x_1}S_{T_1,T_2}(x_1,x_2)\frac{d}{dx_2}S_{T_3}(x_2), & \text{if } 0\le x_1<x_2,\end{cases}$$

which is calculated where the derivatives exist, and depends on the densities of (T_1, T_2) and T_3.

The total mass of the singular component (when $x_1 = x_2 = x$) is given by

$$\alpha=\int_{x=0}^{\infty}S_{T_1,T_2}(x,x)f_{T_3}(x)\mathrm{d}x.$$

First of all, we need to obtain an expression for the joint density $f_{X_1,X_2}(x_1,x_2)$ of the EMO model. It is given in the following

Lemma 6.3 *Let (X_1, X_2) follows the EMO model (6.4). Denote by $f_{T_1,T_2}(x_1,x_2)$ the joint density of (T_1, T_2) and by $f_{T_3}(x)$ the density of T_3. The joint density of EMO model is given by*

$$f_{X_1,X_2}(x_1,x_2)=\begin{cases} g_1(x_1,x_2) & \text{if } x_1>x_2\ge 0;\\ g_0(x) & \text{if } x_1=x_2=x\ge 0;\\ g_2(x_1,x_2) & \text{if } 0\le x_1<x_2,\end{cases}$$

where

$$g_1(x_1,x_2)=f_{T_1,T_2}(x_1,x_2)S_{T_3}(x_1)-\frac{\partial}{\partial x_2}S_{T_1,T_2}(x_1,x_2)f_{T_3}(x_1),$$

$$g_0(x)=f_{T_3}(x)S_{T_1,T_2}(x,x)$$

and

$$g_2(x_1,x_2)=f_{T_1,T_2}(x_1,x_2)S_{T_3}(x_2)-\frac{\partial}{\partial x_1}S_{T_1,T_2}(x_1,x_2)f_{T_3}(x_2).$$

Proof The expressions for $g_1(x_1, x_2)$ and $g_2(x_1, x_2)$ can be obtained by taking partial derivatives $\frac{\partial^2}{\partial x_1\partial x_2}S_{T_1,T_2}(x_1,x_2)$ for $x_1 > x_2$ and $x_1 < x_2$, respectively. We cannot get $g_0(x, x)$ analogously, but from the equation

$$\int_0^{\infty}\int_{x_2}^{\infty}g_1(x_1,x_2)dx_1dx_2+\int_0^{\infty}\int_{x_1}^{\infty}g_2(x_1,x_2)dx_2dx_1+\int_0^{\infty}g_0(x)dx=1.$$

Following the same steps as in Theorem 2.2 in [17], one can explicitly obtain the first two terms in last equation and after some algebra get $g_0(x)$ expression. □

Thus, the general form of the joint density of EMO models is specified by

$$f_{X_1,X_2}(x_1,x_2) = \begin{cases} f_{T_1,T_2}(x_1,x_2)S_{T_3}(x_1) - \frac{\partial}{\partial x_2}S_{T_1,T_2}(x_1,x_2)f_{T_3}(x_1) & \text{if } x_1 > x_2 \geq 0; \\ f_{T_3}(x)S_{T_1,T_2}(x,x) & \text{if } x_1 = x_2 = x \geq 0; \\ f_{T_1,T_2}(x_1,x_2)S_{T_3}(x_2) - \frac{\partial}{\partial x_1}S_{T_1,T_2}(x_1,x_2)f_{T_3}(x_2) & \text{if } 0 \leq x_1 < x_2. \end{cases} \tag{6.12}$$

Note that joint density (6.12) can be written as follows:

$$f_{X_1,X_2}(x_1,x_2) = (1-\alpha)f^{ac}_{X_1,X_2}(x_1,x_2) + \alpha f^{si}_{X_1,X_2}(x_1,x_2),$$

where $\alpha = P(X_1 = X_2) = P(T_3 < \min(T_1,T_2))$ is a constant depending on the parameters of random variables T_1, T_2, and T_3. If $\alpha \in (0,1)$, the absolutely continuous part is given by

$$f^{ac}_{X_1,X_2}(x_1,x_2) = \begin{cases} \frac{g_1(x_1,x_2)}{1-\alpha} & \text{if } x_1 > x_2 \geq 0; \\ \frac{g_2(x_1,x_2)}{1-\alpha} & \text{if } 0 \leq x_1 < x_2, \end{cases}$$

and the singular part is

$$f^{si}_{X_1,X_2}(x_1,x_2) = \begin{cases} \frac{g_0(x)}{\alpha} & \text{if } x_1 = x_2 = x; \\ 0 & \text{otherwise.} \end{cases}$$

We will analyze a football (soccer) dataset of UEFA Champion's League considered by [26], where (i) there was at least one goal scored by the home team, and (ii) there was at least one goal scored directly from a kick (penalty kick, foul kick, or other kick) by any team. Let X_1 be the time (in minutes) of the first kick goal scored by any team, and X_2 be the time of the first goal of any type scored by the home team. With this kind of nonnegative continuous data, one may have three options: $\{X_1 < X_2\}$, $\{X_1 > X_2\}$, or $\{X_1 = X_2\}$, see Fig. 6.3.

We consider five models to analyze UEFA Champion's League dataset: two of them are absolutely continuous and the last three belong to the EMO class.

The absolutely continuous distributions selected for (X_1, X_2) are the Gumbel's (1960) type I bivariate exponential distribution introduced in [11] and the bivariate exponential distribution of Block and Basu (1974), see [6]. The corresponding joint survival functions are given in the third column of Table 6.1, where $\lambda_1, \lambda_2, \lambda_{12} > 0$ and $\theta \in [0,1]$.

To specify the corresponding EMO models, we will assume further that T_3 is exponentially distributed with parameter $\lambda > 0$, i.e., $S_{T_3}(x) = \exp\{-\lambda x\}$ and $f_{T_3}(x) = \lambda\exp\{-\lambda x\}$. Hence, from (6.4) we obtain

$$S_{X_1,X_2}(x_1,x_2) = S_{T_1,T_2}(x_1,x_2)\exp\{-\lambda\max(x_1,x_2)\}.$$

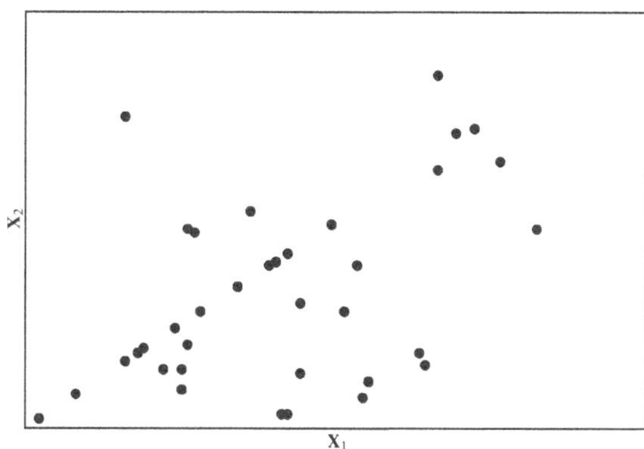

Fig. 6.3 UEFA Champion's League data—Meintanis [26]

Table 6.1 Absolutely continuous models for (X_1, X_2)

Model	Description	$S_{X_1,X_2}(x_1, x_2)$
(M1)	Gumbel [11] type I	$\exp\{-[\lambda_1 x_1 + \lambda_2 x_2 + \theta\lambda_1\lambda_2 x_1 x_2]\}$
(M2)	Block and Basu [6]	$\frac{\lambda_1+\lambda_2+\lambda_{12}}{\lambda_1+\lambda_2}\exp\{-[\lambda_1 x_1 + \lambda_2 x_2 + \lambda_{12}\max(x_1, x_2)]\}$ $-\frac{\lambda_{12}}{\lambda_1+\lambda_2}\exp\{-[(\lambda_1 + \lambda_2 + \lambda_{12})\max(x_1, x_2)]\}$

Table 6.2 EMO models

Model	EMO models	$S_{T_1,T_2}(x_1, x_2)$	$S_{T_3}(\max(x_1, x_2))$
(M3)	MO bivariate model (6.2)	$\exp\{-[\lambda_1 x_1 + \lambda_2 x_2]\}$	$\exp\{-\lambda\max(x_1, x_2)\}$
(M4)	EMO-1 model	Gumbel [11] type I	$\exp\{-\lambda\max(x_1, x_2)\}$
(M5)	EMO-2 model	Block and Basu [6]	$\exp\{-\lambda\max(x_1, x_2)\}$

We choose three versions for $S_{T_1,T_2}(x_1, x_2)$. The corresponding joint survival functions are given in the third column of Table 6.2.

The Bayesian estimation of parameters involved is performed by using *OpenBugs* software (an updated version of WinBugs, consult [22, 41]), which requires the expression for the joint density. In each of our three cases, one can obtain applying (6.12) the joint density $f_{X_1,X_2}(x_1, x_2)$ of EMO models (M3), (M4), and (M5) with $f_{T_3}(\max(x_1, x_2)) = \lambda\exp\{-\lambda\max(x_1, x_2)\}$.

We have to specify the prior distributions for the parameters of the five models in order to start *OpenBugs* procedure. Our choices are given in Table 6.3.

We generated 55,000 samples for the joint posterior distribution of the considered parameters, where the first 5,000 samples are discarded (burn-in sample). Thus we

Table 6.3 Prior distributions of parameters

Parameter	Prior
λ_1, λ_2 and λ_{12}	Gamma (0.1, 0.1)
λ	Gamma (1, 1)
θ	Uniform [0, 1]

eliminated the effect of the initial values for the parameters of the model. Selecting every 10th simulated Gibbs sample, we obtained a final sample of size 5,000 to get the posterior summaries of interest.

The posterior summaries obtained for the five models considered are listed in Tables 6.4, 6.5, 6.6, 6.7, and 6.8, where **Sd** in the third column gives the estimated standard deviation.

The five models to be compared have different number of parameters. To select the "best" model we apply the DIC (Deviance Information Criterion) defined in [40] by

$$DIC = D(\hat{\theta}) + 2n_D = 2\overline{D} - D(\hat{\theta}).$$

In the last relation $D(\hat{\theta})$ is the deviance evaluated at the posterior mean $\hat{\theta} = \mathbb{E}[\theta|\text{data}]$ and $\overline{D} = \mathbb{E}[D(\theta)|\text{data}]$ is the posterior deviance measuring the quality of the data fit for the model. Note that n_D is the effective number of parameters of the model given by $n_D = \overline{D} - D(\hat{\theta})$. Smaller values of DIC indicate better models.

In Table 6.9 we present the resulting models ordered by increasing DIC value criteria. It can be noticed that EMO models (which take into account singularity along the main diagonal) provided a better fit than the absolutely continuous bivariate exponential distributions of [6, 11]. Particularly, the model EMO-2 (considering Block and Basu [6] bivariate exponential distribution for (T_1, T_2)) presented the best fit according to DIC criteria, followed by model EMO-1 (Gumbel [11] type I bivariate exponential distribution for (T_1, T_2)).

We finish this section with a comparison. A Bayesian analysis of the same UEFA Champion's League dataset has been provided by many authors. For example, [18] applied their *Marshall–Olkin bivariate Weibull distribution*, to be denoted as MOBWD, given by

$$S_{X_1,X_2}(x_1, x_2) = \begin{cases} \exp\{-(\lambda_0 + \lambda_1)x_1^{\alpha} - \lambda_2 x_2^{\alpha}\}, & \text{if } x_1 \geq x_2 \\ \exp\{-(\lambda_0 + \lambda_2)x_2^{\alpha} - \lambda_1 x_1^{\alpha}\}, & \text{if } x_2 > x_1. \end{cases}$$

Table 6.4 Model (M1): Gumbel [11] type I bivariate exponential for (X_1, X_2)

Parameter	Mean	Sd	95 % Credible Interval
λ_1	0.0238	0.003964	(0.01663, 0.03216)
λ_2	0.02935	0.004974	(0.02038, 0.03982)
θ	0.1734	0.1588	(0.00450, 0.5925)

Table 6.5 Model (M2): Block and Basu [6] bivariate exponential for (X_1, X_2)

Parameter	Mean	Sd	95 % Credible Interval
λ_1	0.000094	0.000209	(4.42E-08, 0.000669)
λ_2	0.000727	0.001383	(1.06E-06, 0.004847)
λ_{12}	0.045470	0.005589	(0.03553, 0.05784)

Table 6.6 Model (M3): MO bivariate exponential (6.2) for $(T_1, T_2) + T_3 \sim Exp(\lambda)$

Parameter	Mean	Sd	95 % Credible Interval
λ_1	0.005507	0.002521	(0.00164, 0.01134)
λ_2	0.0172	0.00383	(0.01049, 0.02547)
λ	0.01926	0.003882	(0.01224, 0.02749)

Table 6.7 Model (M4): EMO-1 (Gumbel [11] type I for $(T_1, T_2) + T_3 \sim Exp(\lambda)$)

Parameter	Mean	Sd	95 % Credible Interval
λ_1	0.01315	0.003003	(0.007974, 0.01966)
λ_2	0.004985	0.002501	(0.001329, 0.01094)
θ	0.02985	0.005555	(0.01984, 0.02952)
λ	0.3959	0.2801	(0.01376, 0.9522)

Table 6.8 Model (M5): EMO-2 (Block and Basu [6] for $(T_1, T_2) + T_3 \sim Exp(\lambda)$)

Parameter	Mean	Sd	95 % Credible Interval
λ_1	2.20E-04	4.96E-04	(2.04E-03, 0.001639)
λ_2	4.83E-05	1.27E-04	(2.39E-04, 3.75E-01)
λ_{12}	0.0332	0.005764	(0.02273, 0.0448)
λ	0.02034	0.004648	(0.01269, 0.03013)

Table 6.9 Monte Carlo estimates for DIC

Model	Description	DIC
(M5)	EMO-2 (Block and Basu [6] for (T_1, T_2))	546.8
(M4)	EMO-1 (Gumbel [11] type I for (T_1, T_2))	584.0
(M3)	MO bivariate model (6.2)	600.1
(M2)	Block and Basu [6]	633.9
(M1)	Gumbel [11] type I	687.0

In [18] the maximum likelihood estimates (MLE) of parameters α, λ_0, λ_1, and λ_2 in MOBWD and corresponding 95 % confidence intervals (to be abbreviated CI) are reported, see page 279 in their paper. The MLE and their CI can be seen in Table 6.10.

Our MLE and corresponding CI (for the same dataset fitted by the same Marshall–Olkin bivariate Weibull distribution) are listed as well. One can observe a significant difference between results in columns headed "MOBWD" and "Our fits".

In addition, assuming the same Gamma priors with parameters (0.1, 0.1) considered by [18], the corresponding DIC criterion assigns a value 597.1 for the MOBWD. This value is greater in comparison with DIC values of EMO-2 and EMO-1 models being 546.8 and 584.0, respectively. Just consult the last column of Table 6.9.

Table 6.10 MLE and 95 % Credible Interval

Parameter	MOBWD		Our fits	
	MLE	95 % CI	MLE	95 % CI
α	1.6954	(1.3284, 2.0623)	1.6460	(1.0080, 1.8880)
λ_0	2.1927	(1.5001, 2.8754)	0.0081	(0.0008, 0.0285)
λ_1	1.1192	(0.5411, 16973)	0.0042	(0.0004, 0.0125)
λ_2	2.8852	(1.3023, 4.4681)	0.0022	(0.0001, 0.0055)

6.4 A Dual Version of EMO Model

In [9] a dual version of the GMO model (6.3) is introduced and its properties are studied, motivated by its potential applications in risk analysis, financial engineering, and economics. The dual GMO model admits the stochastic representation

$$(Y_1, Y_2) = [\max(D_1, D_3), \max(D_2, D_3)], \tag{6.13}$$

where the continuous random variables D_i, $i = 1, 2, 3$, with support in $\mathbb{R} = (-\infty, \infty)$ are assumed to be independent.

Denote by $F_{Y_1,Y_2}(y_1, y_2) = P(Y_1 \leq y_1, Y_2 \leq y_2)$ the joint distribution function of the random vector (Y_1, Y_2) and by $F_{D_i}(y) = P(D_i \leq y)$ the distribution function of random variables D_i, $i = 1, 2, 3$, where $y \in \mathbb{R}$. Thus, the joint distribution function of the dual GMO model (6.13) is given by

$$\begin{aligned} F_{Y_1,Y_2}(y_1, y_2) &= P(D_1 \leq y_1, D_2 \leq y_2, D_3 \leq \min\{y_1, y_2\}) \\ &= F_{D_1}(y_1) F_{D_2}(y_2) F_{D_3}(\min\{y_1, y_2\}). \end{aligned} \tag{6.14}$$

The model (6.14) is named "dual" in [9] because it is the counterpart of the GMO model based on max instead of min operation in the stochastic representation (6.1). Another difference between GMO model and its dual version is that in the former the random variables T_i, $i = 1, 2, 3$, are assumed to be nonnegative in (6.3).

It happens that the independence assumption between the individual shocks, although adequate in some situations, may not hold in others, as illustrated in the following examples:

1. An insurance company assigns by (6.13) the loss vector of two insured apartments located in the same building. Each insurance policy covers only the largest loss incurred. Both apartments are subject to common disasters, such as earthquakes or tornados, as well as to individual casualties. The common loss can be considered to have the same value for both apartments. Since they are located in the same neighborhood, some individual casualties, such as thefts, are not supposed to be independent;
2. Let (6.13) represents the time to failure vector of two machines operating in the same factory. Each machine can be fed by two different sources of energy: an

individual source and an external one, able to feed both equipment. The machines operate whenever at least one of the sources is properly working. If the two individual sources of energy share some facilities in the factory, their failure times may be dependent random variables.

We extend the dual GMO model assuming dependence between the random variables D_1 and D_2 in (6.13). First, we formalize the dual extended model. After that we list its basic probabilistic properties, obtain the corresponding copula and the distribution of the inactivity times. Several stochastic order comparisons are presented at the end.

6.4.1 Model Specification and Basic Probabilistic Properties

One can define the dual Extended Marshall–Olkin model as follows. Let D_i, $i = 1, 2, 3$, be continuous random variables with support in $\mathbb{R}$ satisfying the stochastic representation (6.13). Assume that the joint distribution function of (D_1, D_2) is given by $F_{D_1,D_2}(y_1, y_2) = P(D_1 \leq y_1, D_2 \leq y_2)$ and suppose that the continuous random variable D_3 is independent of D_1 and D_2. Following (6.13) we obtain the relation

$$F_{Y_1,Y_2}(y_1, y_2) = P(D_1 \leq y_1, D_2 \leq y_2, D_3 \leq \min\{y_1, y_2\}).$$

Thus, we have the following

Definition 6.2 The dual Extended Marshall–Olkin model (to be abbreviated d-EMO) is specified by its joint distribution function

$$F_{Y_1,Y_2}(y_1, y_2) = F_{D_1,D_2}(y_1, y_2) F_{D_3}(\min\{y_1, y_2\}). \tag{6.15}$$

Observe that the dual GMO model (6.14) can be obtained from (6.15) when D_1 and D_2 are independent random variables.

When the random variables D_i are absolutely continuous with density $f_{D_i}(y)$, [4, 15] define the reversed failure rate $m_{D_i}(y)$ and the remaining reversed failure rate $M_{D_i}(y)$ as

$$m_{D_i}(y) = \frac{f_{D_i}(y)}{F_{D_i}(y)} = \frac{d}{dx}[\ln F_{D_i}(y)] \quad \text{and} \quad M_{D_i}(y) = \int_y^\infty m_{D_i}(u)\, du$$

respectively, $i = 1, 2, 3$. Notice that the distribution functions F_{D_i} admit the representation

$$F_{D_i}(y) = \exp\{-M_{D_i}(y)\}, \quad i = 1, 2, 3. \tag{6.16}$$

The bivariate version of (6.16) for the joint distribution function is given by

$$F_{D_1,D_2}(y_1, y_2) = \exp\{-M_{D_1}(y_1) - M_{D_2}(y_2) + A_{D_1,D_2}(y_1, y_2)\}, \tag{6.17}$$

where the function $A_{D_1,D_2}(y_1, y_2)$ is defined by

$$A_{D_1,D_2}(y_1, y_2) = \ln \frac{F_{D_1,D_2}(y_1, y_2)}{F_{D_1}(y_1)F_{D_2}(y_2)}, \tag{6.18}$$

see [35]. In particular, $A_{D_1,D_2}(y_1, y_2) = 0$ if and only if D_1 and D_2 are independent random variables. As an interpretation, $A_{D_1,D_2}(y_1, y_2)$ postulates (describes) the association between random variables D_1 and D_2 free of (i.e., excluding) the marginal contribution into their mutual (genuine) dependence.

The d-EMO model can be equivalently introduced by expression

$$F_{Y_1,Y_2}(y_1, y_2) = \exp\big\{ - M_{D_1}(y_1) - M_{D_2}(y_2) - M_{D_3}\big(\min(y_1, y_2)\big) + A_{D_1,D_2}(y_1, y_2)\big\}. \tag{6.19}$$

Therefore, the knowledge of the distribution of D_3 and joint distribution of D_1 and D_2 given by (6.17) are necessary to obtain the corresponding d-EMO distribution. An important component in (6.17) is the dependence function $A_{D_1,D_2}(y_1, y_2)$, given by (6.18), representing the dependence structure between D_1 and D_2 in addition to the marginal influence.

Using the above notations and relations, we list properties of d-EMO models which are not necessarily exchangeable. The statements are analogous to the corresponding properties of EMO models obtained in Sect. 6.2 and address bounds of joint distribution function, positive and negative dependence properties, corresponding copula, and an aging result.

Lemma 6.4 *The lower and upper bounds for the distribution function of the d-EMO distribution (6.19), are given by*

$$L(y_1, y_2) \le F_{Y_1,Y_2}(y_1, y_2) \le U(y_1, y_2),$$

where

$$L(y_1, y_2) = \max\left\{\exp\left\{-M_{D_1}(y_1)\right\} + \exp\left\{-M_{D_2}(y_2)\right\} - 1, 0\right\} \exp\left\{-M_{D_3}(\min(y_1, y_2))\right\}$$

and

$$U(y_1, y_2) = \min\big\{ \exp\{-M_{D_1}(y_1)\}, \exp\{-M_{D_2}(y_2)\}\big\} \exp\big\{ - M_{D_3}\big(\min(y_1, y_2)\big)\big\}.$$

Note that the bounds $L(y_1, y_2)$ and $U(y_1, y_2)$ in Lemma 6.4 are sharper than the usual Fréchet–Hoeffding bounds, because of the multiplier $\exp\big\{ - M_{D_3}\big(\min (y_1, y_2)\big)\big\}$.

The next statement characterizes the NQD property of the d-EMO model.

Lemma 6.5 *The d-EMO model (6.15) is NQD if and only if*

$$F_{D_1,D_2}(y_1, y_2) \le F_{D_1}(y_1)F_{D_2}(y_2)F_{D_3}(\max\{y_1, y_2\}), \tag{6.20}$$

or equivalently,

$$A_{D_1,D_2}(y_1, y_2) + M_{D_3}\big(\max(y_1, y_2)\big) \leq 0$$

for all $y_1, y_2 \in \mathbf{R}$.

In the sequel we obtain the copula of the d-EMO distribution. By Sklar's theorem, the dependence structure of a bivariate random vector (Y_1, Y_2) with continuous marginal distributions F_{Y_1} and F_{Y_2} can be described by unique copula

$$C_{Y_1,Y_2}(u, v) = F_{Y_1,Y_2}\big(F_{Y_1}^{-1}(u), F_{Y_2}^{-1}(v)\big), \quad (u, v) \in [0, 1],$$

where $F_{Y_i}^{-1}(u)$ is the (generalized) inverse of $F_{Y_i}(y)$, $i = 1, 2$. Observe, that the marginal distribution functions of $F_{Y_1,Y_2}(y_1, y_2)$ from (6.19) are given by $F_{Y_i}(y_i) = \exp\{-M_{D_i}(y_i) - M_{D_3}(y_i)\}$. Denote by $\overline{G}_i(y_i) = M_{D_i}(y_i) + M_{D_3}(y_i)$, $i = 1, 2$. Therefore, we have

Lemma 6.6 *The copula $C_{Y_1,Y_2}(u, v)$ of the d-EMO distribution is given by*

$$C_{Y_1,Y_2}(u, v) = \begin{cases} uv\exp\left\{M_{D_3}(\overline{G}_1^{-1}(-\ln u)) + \overline{G}(u, v)\right\}, & \text{if } \overline{G}_1^{-1}(-\ln u) > \overline{G}_2^{-1}(-\ln v); \\ uv\exp\left\{M_{D_3}(\overline{G}_2^{-1}(-\ln v)) + \overline{G}(u, v)\right\}, & \text{if } \overline{G}_1^{-1}(-\ln u) \leq \overline{G}_2^{-1}(-\ln v), \end{cases} \tag{6.21}$$

where $u, v \in (0, 1]$ and $\overline{G}(u, v) = A_{D_1,D_2}\big(F_{Y_1}^{-1}(u), F_{Y_2}^{-1}(v)\big)$.

Remark 6.7 The function $\exp\{\overline{G}(u, v)\}$ is the only product extra multiplier in (6.21) in addition to the copula expression corresponding to dual GMO distribution and numbered (4) by [9]. This extra term permits NQD modeling on (Y_1, Y_2).

Example 6.3 (Copula of d-EMO distribution under proportional reversed failure rate marginals) Consider a baseline remaining reversed failure rate function $M(y)$ and suppose $F_{D_i}(y) = [\exp\{-M(y)\}]^{\lambda_i}$, $\lambda_i > 0$, $i = 1, 2, 3$, see [12]. After some algebra relation (6.21) simplifies to

$$C_{Y_1,Y_2}(u, v) = \begin{cases} u^{\frac{\lambda_1}{\lambda_1+\lambda_3}} v\exp\{\overline{G}(u, v)\}, & \text{if } \overline{G}_1^{-1}(-\ln u) > \overline{G}_2^{-1}(-\ln v); \\ uv^{\frac{\lambda_2}{\lambda_2+\lambda_3}}\exp\{\overline{G}(u, v)\}, & \text{if } \overline{G}_1^{-1}(-\ln u) \leq \overline{G}_2^{-1}(-\ln v), \end{cases}$$

where $u, v \in (0, 1]$. Notice that when D_1 and D_2 are independent we have $\overline{G}(u, v) = 0$ for all $u, v \in (0, 1]$ and we obtain the survival copula of the Marshall–Olkin bivariate exponential distribution, see the corresponding comments in [20].

Let us mention a specific aging property of a bivariate random vector (Y_1, Y_2) that follows the d-EMO model (6.19). Denote by

$$\mathbf{Y}_{(t)} = \big[(t - Y_1, t - Y_2) \mid Y_1 \leq t, Y_2 \leq t\big], \quad t \geq 0$$

the corresponding *inactivity times* vector which has nonnegative marginals. Consider the vector

$$-\mathbf{Y}_{(t)} = \left[(Y_1 - t, Y_2 - t) \mid Y_1 \le t, Y_2 \le t\right], \tag{6.22}$$

which is the symmetric image of $\mathbf{Y}_{(t)}$ with respect to the point (0, 0). We are interested to examine the distribution of the random vector $-\mathbf{Y}_{(t)}$. The result is given in the following

Lemma 6.7 *If the vector* (Y_1, Y_2) *follows a d-EMO distribution, then so does* $-\mathbf{Y}_{(t)}$ *for any* $t \ge 0$.

As a consequence, we have the following

Corollary 6.1 *If the vector* (Y_1, Y_2) *follows a d-EMO distribution, then* (Y_1, Y_2) *and* $-\mathbf{Y}_{(t)}$ *have the same type of copula, given by relation (6.21).*

6.4.2 Stochastic Order Comparisons

We present several stochastic order comparisons between bivariate random vectors that follow a d-EMO model. The first result is related to the usual stochastic order between random vectors (Y_1, Y_2) and (Z_1, Z_2), to be denoted by $(Y_1, Y_2) \le_{st} (Z_1, Z_2)$. This means that $\mathrm{E}[\psi(Y_1, Y_2)] \le \mathrm{E}[\psi(Z_1, Z_2)]$ for every increasing function ψ such that the expectation exists, see [36].

Lemma 6.8 *Consider continuous random variables* V_i *and* D_i, $i = 1, 2, 3$, *and suppose* $(V_1, V_2) \le_{st} (D_1, D_2)$ *and* $V_3 \le_{st} D_3$. *Then* $(Y_1, Y_2) \le_{st} (Z_1, Z_2)$, *where*

$$\begin{aligned}(Y_1, Y_2) &= [\max(V_1, V_3), \max(V_2, V_3)] \text{ and } (Z_1, Z_2)\\ &= [\max(D_1, D_3), \max(D_2, D_3)]\end{aligned}$$

follow d-EMO distribution.

Proof Since V_3 is independent of (V_1, V_2) and D_3 is independent of (D_1, D_2), we have $(V_1, V_2, V_3) \le_{st} (D_1, D_2, D_3)$. Taking into account that $\max(y, z)$ is increasing in y and z, the result follows from Theorem 6.B.16(a) from [36]. □

Remark 6.8 Notice that the function $g(v_1, v_2, v_3) = [\max(v_1, v_3), \max(v_2, v_3)]$ is increasing and convex in its arguments. Thus, Lemma 6.8 remains valid if we replace the usual stochastic order by the increasing convex order, as a direct application of Theorem 7.A.5(a) from [36].

The concordance order for copulas of d-EMO distributions is considered in the sequel. In Lemma 6.9 we compare the copulas of d-EMO distributions obtained from the same pair of random variables (V_1, V_2) but considering two different common shocks represented by random variables V_3 and D.

Lemma 6.9 *Suppose V_i, $i = 1, 2, 3$, and D are continuous random variables and $A_{V_1,V_2}(y_1, y_2) = \ln \frac{F_{V_1,V_2}(y_1,y_2)}{F_{V_1}(y_1)F_{V_2}(y_2)}$ is nondecreasing in y_1 and y_2. If $V_3 \geq_{st} D$, then $C_{Y_1,Y_2}(u, v) \geq C_{Z_1,Z_2}(u, v)$, for all $u, v \in [0, 1]$, where*

$$(Y_1, Y_2) = [\max(V_1, V_3), \max(V_2, V_3)] \text{ and } (Z_1, Z_2) = [\max(V_1, D), \max(V_2, D)]$$

follow d-EMO distributions.

In Lemma 6.10 we consider the case where two d-EMO distributions are composed by the same common shock (represented by a random variable V_3) and different individual shocks, represented by the pairs of random variables (V_1, V_2) and (D_1, D_2).

Lemma 6.10 *Suppose V_i, $i = 1, 2, 3$, and D_i, $i = 1, 2$, are continuous random variables and let $A_{V_1,V_2}(y_1, y_2) = \ln \frac{F_{V_1,V_2}(y_1,y_2)}{F_{V_1}(y_1)F_{V_2}(y_2)} \geq A_{D_1,D_2}(z_1, z_2) = \ln \frac{F_{D_1,D_2}(z_1,z_2)}{F_{D_1}(z_1)F_{D_2}(z_2)}$ for $y_i \leq z_i$, $i = 1, 2$. Define the d-EMO distributions*

$$(Y_1, Y_2) = [\max(V_1, V_3), \max(V_2, V_3)] \text{ and } (Z_1, Z_2) = [\max(D_1, V_3), \max(D_2, V_3)].$$

If $V_1 \leq_{st} D_1$, $V_2 \leq_{st} D_2$, then $C_{Y_1,Y_2}(u, v) \geq C_{Z_1,Z_2}(u, v)$, for all $u, v \in [0, 1]$. In addition we have $(Y_1, Y_2) \leq_{st} (Z_1, Z_2)$.

Example 6.4 (Insurance application) Consider an insurance company offering two policies. The first policy covers a pair of random losses (Y_1, Y_2) which are subject to dependent individual risks (with potential losses V_1 and V_2). The second one covers the random losses (Z_1, Z_2), also exposed to dependent individual risks denoted by D_1 and D_2. Moreover, there exists a common risk, with random loss V. If the policies cover only the largest loss, we have

$$(Y_1, Y_2) = [\max(V_1, V), \max(V_2, V)] \quad \text{and} \quad (Z_1, Z_2) = [\max(D_1, V), \max(D_2, V)].$$

Suppose that for each of these two policies, the retained loss of the insurance company is some nondecreasing function $h_i(.)$ of the random losses, $i = 1, 2$. For the first policy its expected retained loss is given by $\mathrm{E}[h_1(Y_1) + h_2(Y_2)]$. Analogously, $\mathrm{E}[h_1(Z_1) + h_2(Z_2)]$ is the expected retained loss for the second policy.

Let the joint distribution of (V_1, V_2) and (D_1, D_2) be given by

$$F_{V_1,V_2}(y_1, y_2) = \frac{(1-\exp\{-\lambda_1 y_1\})(1-\exp\{-\lambda_1 y_2\})}{1-\theta_1\exp\{-\lambda_1 y_1 - \lambda_1 y_2\}}, \quad \theta_1 \in [0, 1]$$

and

$$F_{D_1,D_2}(y_1, y_2) = \frac{(1-\exp\{-\lambda_2 y_1\})(1-\exp\{-\lambda_2 y_2\})}{1-\theta_2\exp\{-\lambda_2 y_1 - \lambda_2 y_2\}}, \quad \theta_2 \in [-1, 0],$$

where $\lambda_1 \geq \lambda_2 > 0$ and $y_i \geq 0,\ i = 1, 2$. Then the conditions of Lemma 6.10 hold true and we conclude that $(Y_1, Y_2) \leq_{st} (Z_1, Z_2)$. Thus, we have

$$\mathrm{E}[h_1(Y_1) + h_2(Y_2)] \leq \mathrm{E}[h_1(Z_1) + h_2(Z_2)],$$

provided the expectations exist. Therefore, the expected retained loss is smaller in the first insurance policy than in the second one.

With additional assumptions in Lemma 6.10, we can obtain another related concordance order result for copulas of d-EMO distributions, see Chap. 4 in [32].

6.5 Concluding Remarks

The classical bivariate exponential MO distribution (6.2) finds applications in reliability, survival analysis, finance and life insurance, among other fields. Hence, any extension of this model has its theoretical and practical relevance. We extended the classical MO model by assuming dependence between the individual shocks represented by the random variables T_1 and T_2 in the stochastic representation (6.1). The dependence structure in EMO model defined by (6.4) or (6.7) enlarges the field of applications of MO and GMO distributions.

In Sect. 6.2 we examined some probabilistic properties of EMO model. We obtained its survival copula representation and investigated the distributional property of residual lifetimes. In spite of the bivariate lack-of-memory property (LMP) does not hold true for all EMO models, it was shown that the original vector $(\min(T_1, T_3), \min(T_2, T_3))$ and the corresponding vector of residual lifetimes $\mathbf{X}_t$ have the same type of survival copula given by (6.9). In [8] a similar relation is established for the so-called almost bivariate LMP survival copula.

A deep study of EMO models is presented in [32], Chap. 3. An absolutely continuous version of EMO model that preserves the stochastic representation (6.1) was obtained. Weak bivariate aging properties were analyzed for the absolutely continuous version of the EMO model. In addition, the bivariate LMP of EMO models were investigated and only EMO models with singular component have this property. It was shown that if $S_{T_1,T_2}(x_1, x_2)$ is absolutely continuous then (X_1, X_2) possesses the bivariate LMP with exponential marginals if and only if (T_1, T_2) possess the bivariate LMP and T_i are exponentially distributed, $i = 1, 2, 3$. Extreme value analysis of EMO models is presented in [32] as well. It was shown that if $S_{T_1,T_2}(x_1, x_2)$ is absolutely continuous then extreme value EMO models with exponential marginals can be obtained if and only if the survival copula that joins (T_1, T_2) is an extreme value copula and T_i are exponentially distributed, $i = 1, 2, 3$.

In Sect. 6.3 we provided a Bayesian analysis of a dataset which displays a singular component along the line $(x_1 = x_2 = x \geq 0)$. Due to this singularity, EMO distributions are suitable and, as expected, exhibited a better performance than the two absolutely continuous models considered. For comparison purposes, in the three EMO distributions used in the analysis we fixed the independent common shock T_3 to be exponentially distributed. Notice that in the MO bivariate exponential distribution (6.2) the only source of dependence between the observed random variables X_1 and X_2 comes from this common shock. From Table 6.9 it can be seen that among the models with singular component, the MO distribution (6.2) has the weakest performance. The additional dependence between the individual shocks T_1 and T_2 introduced in EMO-1 and EMO-2 models has the effect of providing a better fit to the data. It is worth noting that the parameters of EMO models can be also estimated by the classical maximum likelihood procedure using the density function (6.12). Meanwhile, the complexity of the maximization procedure heavily depends on the functional form of the survival function $S_{T_1,T_2}(x_1, x_2)$.

In Sect. 6.4 we extended the model introduced in [9] relaxing authors assumption of independence between the individual shocks. In the proposed dual extended Marshall–Olkin distribution (6.19), the dependence structure between its two components is explained not only by the common shock, but also by the joint distribution of the individual shocks. Probabilistic properties, copula representation, distribution of the inactivity times, and stochastic order comparisons are presented for the dual extended model.

A possible application of EMO-type distributions is related to the following finance scenario. Let us consider financial institutions holding notes and Treasury bonds, issued by governments, as part of their investment and risk management policies. Governments, by their turn, concerned with defaults in financial system and their negative impact in the economy, implicitly or explicitly provide insurance to banking sector, at the cost of weakening their own balance sheets. This relationship makes default of government and financial system-dependent events, whose probability of occurrence is particularly important to be assessed during financial crisis. Unfortunately, since it is possible to observe a joint default (for instance, the one observed in the recent Iceland crisis), the use of conventional MO-type models should be avoided.

The extension of the Marshall–Olkin framework to allow for dependence in the shocks specific to the issuers (identified by the dependence between T_1 and T_2 in (6.1)), and between each of these and the common factor representing systemic risk (e.g., by T_3) is paramount from the point of view of applications. In fact, there is a crucial econometric problem that these models can resolve. On one side it is not reasonable to assume that issuer specific shocks are fully independent, so that all the dependence in the system could be attributed to the common factor. On the other side, ignoring these relationships may induce a substantial bias in the estimates of the common factor itself. This problem is recognized by [1], who propose an estimator based on a model with standard bivariate Marshall–Olkin margins.

We find a potential application of EMO and d-EMO classes in commodity and energy markets modeling as well, by assuming a dependent structure between error terms of individual time series (identified by T_1 and T_2), see a related discussion in [13].

We do believe that EMO and d-EMO models will be both of further theoretical and practical interests. Several recent investigations related to the newly introduced concept of Sibuya-type bivariate LMP and its invariant copula representation can be found in [33, 34]. Multivariate extensions, related statistical inference and appropriate applications are possible objects of future research.

Acknowledgments The authors are grateful for the precise referee suggestions which highly improved earlier versions of the manuscript. The first author acknowledges the sponsorship of Central Bank of Brazil under the Graduate Program PPG. The second author is partially supported by FAPESP (2013/07375-0 and 2011/51305-0) and CNPq grants. We are thankful to Leandro Ferreira for the help with OpenBugs software.

References

1. Baglioni, A., Cherubini, U.: Within and between systemic country risk. Theory and evidence from the sovereign crisis in Europe. J. Econ. Dyn. Control **37**, 1581–1597 (2013)
2. Balakrishnan, N., Lai, C.-D.: Continuous Bivariate Distributions, 2nd edn. Springer, New York (2009)
3. Barlow, R., Proschan, F.: Statistical Theory of Reliability and Life Testing. Silver Spring, Maryland (1981)
4. Barlow, R., Marshall, A., Proschan, F.: Properties of probability distributions with monotone hazard rate. Ann. Math. Stat. **34**, 375–389 (1963)
5. Bassan, B., Spizzichino, F.: Relations among univariate aging, bivariate aging and dependence for exchangeable lifetimes. J. Multivar. Anal. **93**, 313–339 (2005)
6. Block, H., Basu, A.: A continuous bivariate exponential extension. J. Am. Stat. Assoc. **69**, 1031–1037 (1974)
7. Dabrowska, D.: Kaplan-Meier estimate on the plane. Ann. Stat. **16**, 1475–1489 (1988)
8. Dimitrov, B., Kolev, N.: The BALM copula. Int. J. Stoch. Anal. **2013**, 1–6 (2013)
9. Fang, R., Li, X.: A note on bivariate dual generalized Marshall-Olkin distributions with applications. Probab. Eng. Inf. Sci. **27**, 367–374 (2013)
10. Freund, J.: A bivariate extension of the exponential distribution. J. Am. Stat. Assoc. **56**, 971–977 (1961)
11. Gumbel, E.: Bivariate exponential distributions. J. Am. Stat. Assoc. **55**, 698–707 (1960)
12. Gupta, R.C., Gupta, R.D., Gupta, P.L.: Modeling failure time data by Lehman alternatives. Commun. Stat. Theory Methods **27**, 887–904 (1988)
13. Jaschke, S., Siburg, K., Stoimenov, P.: Modelling dependence of extreme events in energy markets using tail copulas. J. Energy Mark. **5**(4), 63–80 (2013)
14. Karlis, D.: ML estimation for multivariate shock models via an EM algorithm. Ann. Inst. Stat. Math. **55**, 817–830 (2003)
15. Keilson, J., Sumita, U.: Uniform stochastic ordering and related inequalities. Can. J. Stat. **10**, 181–198 (1982)

16. Kundu, D., Dey, A.: Estimating the parameters of the Marshall-Olkin bivariate Weibull distribution by EM algorithm. Comput. Stat. Data Anal. **53**, 956–965 (2009)
17. Kundu, D., Gupta, R.: Modified Sarhan-Balakrishnan singular bivariate distribution. J. Stat. Plan. Inference **140**, 525–538 (2010)
18. Kundu, D., Gupta, A.: Bayes estimation for the Marshall-Olkin bivariate Weibull distribution. Comput. Stat. Data Anal. **57**, 271–281 (2013)
19. Lehmann, E.: Some concepts of dependence. Ann. Math. Stat. **37**, 1137–1153 (1966)
20. Li, X.: Duality of the multivariate distributions of Marshall-Olkin type and tail dependence. Commun. Stat. Theory Methods **37**, 1721–1733 (2008)
21. Li, X., Pellerey, F.: Generalized Marshall-Olkin distributions and related bivariate aging properties. J. Multivar. Anal. **102**, 1399–1409 (2011)
22. Lunn, D., Spiegelhalter, D., Thomas, A., Best, N.: The BUGS project: evolution, critique and future directions (with discussion). Stat. Med. **28**, 3049–3082 (2009)
23. Mai, J.-F., Scherer, M.: Simulating Copulas. Imperial College Press, London (2012)
24. Marshall, A., Olkin, I.: A multivariate exponential distribution. J. Am. Stat. Assoc. **62**, 30–41 (1967)
25. McNeil, A., Frey, L., Embrechts, P.: Quantitative Risk Management. Princeton University Press, Princeton (2005)
26. Meintanis, S.: Test of fit for Marshall-Olkin distributions with applications. J. Stat. Plan. Inference **137**, 3954–3963 (2007)
27. Mirhosseini, S., Amini, M., Kundu, D., Dolati, A.: On a new absolutely continuous bivariate generalized exponential distribution. To appear in Statistical Methods and Applications (2015)
28. Mohsin, M., Kazianka, H., Pilz, J., Gebhardt, A.: A new bivariate exponential distribution for modeling moderately negative dependence. Stat. Methods Appl. **23**, 123–148 (2014)
29. Mulero, J., Pellerey, F.: Bivariate aging properties under Archimedean dependence structures. Commun. Stat. Theory Methods **39**, 3108–3121 (2010)
30. Muliere, P., Scarsini, M.: Generalization of Marshall-Olkin type of distributions. Ann. Inst. Stat. Math. **39**, 429–441 (1987)
31. Nelsen, R.: An Introduction to Copulas, 2nd edn. Springer, New York (2006)
32. Pinto, J.: Deepening the notions of dependence and aging in bivariate probability distributions. Ph.D. thesis, Sao Paulo University (2014)
33. Pinto, J., Kolev, N.: Sibuya-type bivariate lack of memory property. J. Multivar. Anal. **134**, 119–128 (2015)
34. Pinto, J., Kolev, N.: Copula representations for invariant dependence functions. In: Glau, K., Scherer, M., Zagst, R. (eds.) Innovations in Quantitative Risk Management. Springer Series in Probability & Statistics, vol. 99, pp. 411–421. Springer, New York (2015)
35. Sankaran, P.G., Gleeja, V.L.: On bivariate reversed hazard rates. J. Jpn. Stat. Soc. **36**, 213–224 (2006)
36. Shaked, M., Shanthikumar, J.: Stochastic Orders. Springer, New York (2007)
37. Shoaee, S., Khorram, E.: A new absolute continuous bivariate generalized exponential distribution. J. Stat. Plan. Inference **142**, 2203–2220 (2012)
38. Sibuya, M.: Bivariate extreme statistics I. Ann. Inst. Stat. Math. **11**, 195–210 (1960)
39. Singpurwalla, N.: Reliability Analysis: A Bayesian Perspective. Wiley, Chichester (2006)
40. Spiegelhalter, D., Best, N., Carlin, B., van der Linde, A.: Bayesian measures of model complexity and fit (with discussion). J. R. Stat. Soc. Ser. A **64**, 583–639 (2002)
41. Spiegelhalter, D., Thomas, A., Best, N., Lunn, D.: WinBugs Version 1.4, User Manual. Institute of Public Health and Department of Epidemiology and Public Health. Imperial College School of Medicine, London (2003)

Zeitfracht Medien GmbH
Ferdinand-Jühlke-Straße 7
99095 Erfurt, Deutschland
produktsicherheit@kolibri360.de